System Innovation for Sustainability 4

System Innovation for Sustainability 4

CASE STUDIES IN SUSTAINABLE CONSUMPTION AND
PRODUCTION — ENERGY USE AND THE BUILT ENVIRONMENT

EDITED BY SAADI LAHLOU

Greenleaf
PUBLISHING

© 2011 Greenleaf Publishing Limited

Published by Greenleaf Publishing Limited
Aizlewood's Mill
Nursery Street
Sheffield S3 8GG
UK
www.greenleaf-publishing.com

Printed in Great Britain on acid-free paper by
CPI Antony Rowe, Chippenham and Eastbourne

Cover by LaliAbril.com

British Library Cataloguing in Publication Data:
   A catalogue record for this book is available from the British Library.

   ISBN-13: 9781906093259

Contents

1
Introduction

Saadi Lahlou
EDF R&D, Clamart, France; London School of Economics and Political Science, London UK

Martin Charter and Tim Woolman
The Centre for Sustainable Design, University for the Creative Arts, Farnham, UK

Arnold Tukker
TNO Built Environment and Geosciences, The Netherlands;
Norwegian University of Science and Technology (NTNU), Department of Product Design

This publication is a result of the European project SCORE! (Sustainable Consumption Research Exchange) and summarises the findings of the Working Group on Sustainable Consumption and Production (SCP) in the housing domain, focusing on energy use.

Housing is, with transport, a principal consumption domain for energy. According to the Environmental Impact of Products (EIPRO) report (Tukker *et al.* 2006), housing accounts for about a quarter of the environmental impact from the general consumption of products in the European Union, on a par with agriculture, food and mobility. These three consumption domains were selected in the SCORE! project as the areas to be tackled through case-study research and expert exchange to understand how a shift to more sustainable consumption and production can be organised and how implementation can be supported.

Energy consumption in housing and buildings is a key issue for sustainability, primarily because it contributes to the depletion of non-renewable fossil fuels and the production of carbon dioxide (CO_2) and other pollution. Trends in energy supply and demand can affect growth and indirectly affect many aspects of human activity. The opportunities to apply research to 'windows of opportunity' and to influence energy consumption in housing and buildings are therefore a major target for SCP policies, promoting implementation at the macro, meso and micro levels through a range of stakeholders. A growing body of evidence shows that cases demonstrating action towards SCP in energy use in housing can inspire innovation through a range of actors. Within the framework of the SCORE! project, our aim is to inform and accelerate steps towards sustainable, systemic change.

1.1 The SCORE! project

SCORE! is a network initiated with EU funding supporting the development of the UN 10 Year Framework of Programmes on Sustainable Consumption and Production. The mission of SCORE! has been to organise a leading science network that provides input to this framework. The EU funding for SCORE! ran between 2005 and 2008, engaging 28 institutions. By the organisation of major workshops and conferences the project engaged and structured a larger community of a few hundred professionals in the EU and beyond

The SCORE! philosophy assumes that SCP structures can be realised only if experts that understand business development, (sustainable) solution design, consumer behaviour and system innovation policy work together in shaping those structures. Furthermore, this should be linked with experiences of actors (industry, consumer groups, eco-labelling organisations) in real-life consumption areas: mobility, agro-food, and energy use in housing—together responsible for 70% of the life-cycle environmental impacts of Western societies (Hertwich 2005; Tukker et al. 2006). This led to the following approach to the project in two phases.

The first phase of the project, marked by a workshop organised with the support of the European Environment Agency (EEA) in Copenhagen in April 2006, aimed to arrange a positive confrontation of conceptual insights developed in the four aforementioned science communities (business development, sustainable solution design, consumer behaviour and system innovation) of how 'radical' change to SCP can be governed and realised. The results of this phase were published as the first book in the 'System Innovation for Sustainability' series (Tukker et al. 2008)

The second phase put the three consumption areas at centre stage. SCORE! Work package leaders inventoried cases that work with examples of successful switches to SCP in their field. In a series of conferences and workshops, these cases were analysed for 'implementability', adapted where needed, and policy suggestions were worked out that can support implementation. The results of this phase are published in this book and in two parallel books covering the areas of mobility, and agriculture and food (Geerken and Borup 2009; Tischner et al. 2010). The Paris SCORE! conference produced reviewed case studies (Lahlou and Emmert 2007), a selection of which are presented in this book.

As the SCORE! project will be followed up and as activities will be continued the organisers hope for further fruitful cooperation with the established group of experts as well as with new members. The reader is recommended to visit the SCORE! website for information on the continuation of activities contributing to SCP research in Europe and internationally.

1.2 Process, contributors and responsibilities for work on energy use in housing

1.2.1 Process

As the SCORE! project was a Coordinated Action within the EU 6th Framework research programme, the aim was to exchange and disseminate good practice and to promote and support the networking and coordination of research and innovation activities. Having built up a research community in the field of SCP, including experts in the different science fields and consumption domains, the two SCORE! project conferences and three interactive workshops enabled experts to focus on three consumption domains: energy use in housing (focusing on energy use), mobility and agriculture and food.

During the SCORE! conferences and workshops dedicated sessions on energy use in houses and buildings were held. Researchers sent in cases that showed either success or failure on the path to sustainable energy use in housing. During the conferences and workshops, cases that demonstrated learning from implementation were selected from a diverse inventory and the lessons learned were refined through discussion. This enabled conclusions and proposed steps for transition to more sustainable energy use in housing to be developed. Selected, edited cases presented in these sessions on energy use in housing have been included in the case study part of this book to provide more specific evidence. The proceedings of these sessions at the conferences are available on the SCORE! project website (www.score-network.org).

In the final workshop and conference, the team responsible for the area of energy use in housing presented thought-provoking statements and draft conclusions to stimulate debate on problems, trends, windows of opportunity and actions in a specific session that fed into the concluding chapter of this book.

1.2.2 Contributors and responsibilities

The SCORE! team members from Electricité de France (EDF) and The Centre for Sustainable Design (CfSD) were responsible for running the process within the area of energy use in housing. They organised the interaction during the workshops, performing desk-based research, writing an initial system analysis of the need area and providing conclusions across the cases presented.

The SCORE! network members participating in the project contributed by providing and presenting case studies and by active participation during workshop discussions. Also, other participants from the research community, policy circles, industry and non-governmental organisations (NGOs) provided and presented cases and actively participated. We want to thank them all for their input and intense collaboration, providing collective insights in this complex field within SCP.

1.3 Book outline

Chapter 2 provides a generic analysis of the housing and energy use domain, including sustainability challenges and a general analysis of potential opportunities for change. Examples of steps towards sustainable use of energy in houses and buildings, from 'local experiments', through 'innovative communities', to wider regime or non-local scale change, are provided through eight cases presented in the Chapters 3–10.

In Chapter 3 Brown and Vergragt illustrate the crucial importance of the quality of vision, project leaders and teams in the success of innovation at the first level of local experimentation towards SCP through the realisation of a design for a 'zero-energy' residential building in Boston.

In Chapter 4 Wimmer and Kang trace the history of a project creating a sustainable house made from straw bales; the S-House achieves a Factor 10 improvement in resource efficiency and provides an ongoing demonstration to disseminate the constituent technical innovations.

In Chapter 5 Wüstenhagen describes how a series of solar houses in Freiburg were built, examining the role of a 'champion', Rolf Disch, who became a social entrepreneur in creating social real estate funds to implement a new generation of houses that produce more energy than they consume.

In Chapter 6 Thorp describes how Woking Borough Council in the UK strives to reduce energy-related emissions from houses and buildings exceeding environmental policy targets. It has done this through holistic governance and technical and commercial innovation as well as by establishing an energy services company, enabling large novel fuel cell, combined heat and power (CHP) and photovoltaic (PV) installations and private-wire local distribution.

In Chapter 7 Kaltenegger and Tisch identified and analysed barriers linked to the implementation of product–service systems (PSS) in public procurement in Austria and strategies to deal with them in the context of energy-saving performance contracting for federally owned public buildings in Austria.

In Chapter 8 Loftness *et al.* address the renovation market issue through a tool developed at Carnegie Mellon University—BIDS™ (Building Investment Decision System)—addressing a block in the sustainable decision chain by feeding decision-makers with life-cycle performance and return on investment (ROI) of energy-saving architectural options based on actual cases.

In Chapter 9 Fischer reviews the influence of consumer feedback to stimulate electricity conservation through an analysis of 26 case studies, clearly showing that detailed metering and consumer feedback is a necessary step towards sustainability.

In Chapter 10 Reusswig, Lorek and Fuchs analyse the history of wind energy generation in Germany, showing how the influences of citizens, consumers and different regulation frameworks enable a gradual structuring of wind energy production and consumption into a set of interconnected politico-economic, social and lifestyle changes combined in a slow transitional process.

In Chapter 11 Lahlou *et al.* offer conclusions, reflecting on the findings and policy implications of the preceding chapters and suggesting how stakeholders can find a path to SCP in energy use in housing, notably through an appreciation of the three 'layers' influencing systemic change: physical affordances (from the building and from equipment), popular representations and practice and the institutional 'rules of the game'.

References

Geerken, T., and M. Borup (eds.) (2009) *System Innovation for Sustainability 2: Case Studies in Sustainable Consumption and Production — Mobility* (Sheffield, UK: Greenleaf Publishing; www.greenleaf-publishing.com/scp2).

Hertwich, E. (2005) 'Life-cycle Approaches to Sustainable Consumption: A Critical Review', *Environmental Science and Technology* 39.13: 4673.

Lahlou, S., and S. Emmert (eds.) (2007) 'Sustainable Consumption and Production Cases in the Domain of Food, Mobility and Housing', in *Proceedings of the Workshop of the Sustainable Consumption Research Exchange (SCORE) Network, 4 and 5 June 2007, Paris, France*; www.score-network.org/score/score_module/index.php?cat_name=cat_t_sco_milestonedoc&mst_id=23 (accessed October 2009).

Tischner, U., E. Stø, U. Kjærnes and A. Tukker (eds.) (2010) *System Innovation for Sustainability 3: Case Studies in Sustainable Consumption and Production — Food and Agriculture* (Sheffield, UK: Greenleaf Publishing; www.greenleaf-publishing.com/scp3).

Tukker, A., M. Charter, C. Vezzoli, E. Stø and M. Munch Andersen (eds.) (2008) *System Innovation for Sustainability 1: Perspectives on Radical Change to Sustainable Consumption and Production* (Sheffield, UK: Greenleaf Publishing; www.greenleaf-publishing.com/scp1).

——, G. Huppes, S. Suh, R. Heijungs, J. Guinee, A. de Koning, T. Geerken, B. Jansen, M. van Holderbeke and P. Nielsen (2006) *Environmental Impacts of Products* (Seville, Spain: European Science and Technology Observatory/Institute for Prospective Technological Studies [ESTO/IPTS]).

2

Energy use in houses and buildings and sustainable consumption

Saadi Lahlou
EDF R&D, France; London School of Economics and Political Science, UK

Martin Charter and Tim Woolman
The Centre for Sustainable Design, UK

Energy consumption is obviously a key issue for sustainability, primarily because it depletes non-renewable fossil fuels, produces carbon dioxide (CO_2) and other pollution. As climate change is becoming a key political issue, and as oil prices rise, society has become acutely aware of this issue. Also, as a limited resource, energy can become a major source of conflict (for example in the Middle East), which is another direct threat to sustainability.

Energy is a special issue because it is a key input to almost all other consumptions and production processes. Energy is therefore a crucial parameter controlling growth and determining many aspects of human activity in general. Modifications in energy demand or supply will cause indirect impacts in many other domains. For example, raised prices for energy tend to make activities more local through raising the costs of transportation.

Domains that are big consumers of energy are therefore a major target for sustainability policies; and among those domains housing comes first. Housing is, with transport, the main consumption domain for energy. According to the Environmental Impact of Products (EIPRO) report (Tukker *et al.* 2006), housing accounts for about 25% of the environmental impact from the general consumption of products in the European Union, on a par with food and transport.

Energy use in houses and other buildings is a significant proportion of energy consumption, set to rise with increases in population and the number of associated buildings, notably houses. In France, for example, energy consumption in houses and offices in 2009 accounted for 43% of total national energy consumption (Commissariat Général au Développement Durable 2009: 10), and 23% of national greenhouse gas emissions. The UK's 21 million homes consume around 50 million tonnes of oil equivalent (responsible for 27% of the UK's CO_2 emissions), this energy use having increased steadily, by about 1.3 % per year, since 1990 (UK DTI 2006). Germany's buildings, which consume about half of the energy (Noeren 2007) also contribute about 20% of the country's CO_2 emissions.

Beyond this, buildings are the environment where the majority of us spend most of our lives; they deeply influence many other consumption patterns and are an important factor in our life and comfort. The societal function and nature of buildings as they are currently culturally constructed accounts for many of the difficulties in moving towards sustainable consumption and production (SCP), both present and in the future.

Buildings have a long lifetime; therefore this domain is a major target for any structural change in consumption patterns. Conversely, long lifetimes come with significant 'inertia', therefore the stock of existing buildings is often an obstacle to policies toward behavioural change.

In this chapter we will focus mainly on the use of energy in housing and leave aside the construction aspects.[1] A vast amount of energy is used in housing operation and maintenance—the so-called use phase. This specifically relates to:

● Energy for control of ambiance: lighting, heating, ventilation and air conditioning (HVAC)[2] and cleaning

● Energy for internal activity: cooking and cold storage, washing, entertainment, communication, etc., as well as work in commercial buildings

Many issues are common to homes and commercial buildings (public buildings, offices and industrial and commercial buildings) but the decision systems and actors are somewhat different. In commercial buildings 'evolution' is influenced mainly by financial and technological decision-making by finance directors, architects and facility managers whereas domestic energy consumption patterns are a matter of lifestyle and individual end-user decisions. Still, commercial and domestic buildings partly share the same ecology of providers and professionals (from energy suppliers to appliance manufacturers and architects). And commercial buildings are often the place where future mass-market technologies are first adopted. They should therefore stay in the frame of the system considered. There is perhaps more of a need to take an integrated systems view of SCP in the case of buildings, with the goal of reducing energy consumed in the use phase, as use of this perspective may produce more optimal and smarter solutions.

The system of energy use in houses and buildings will be described in the next section, followed by the sustainability problems and windows of opportunity.

---

1 Construction aspects are important energy wise, and construction waste when buildings are dismantled is also a 'hot' issue. We focus here, though, on aspects where end-users are directly involved.

2 The term HVAC is generally used to refer to integrated systems, but for the sake of simplicity here we will subsume under it all forms of heating and cooling and, more generally, climate control (heaters, heat pumps, etc.) since in the end good thermodynamic monitoring must take into account all these parameters.

2.1 Systemic description of the domain

The stakeholders in the domain of energy consumption in houses and buildings are many, and a description of the regime in detail would go beyond the general scope of this chapter. The problem lies in three areas: the inertia of the building stock, the complexity of the system and the slow rate of change in users' habits. In simple terms, it is easier to make buildings more sustainable in new construction. A building is a system; the nature of the envelope (the structure, such as the walls) of the building will have a strong influence on its energy consumption. Therefore making sustainable retrofits is both costly and difficult. However, as we shall see, new construction is only a minor part of the market. On the user side a great number of small routines need to be changed to create more conservative behaviour.

The current building and housing regime is the result of long-term societal construction that has brought in many legal frameworks and technical regulations connected with security and property issues. The domain can be roughly divided into two sub-domains: residential and commercial buildings. In residential buildings, collective habitats and individual houses are very different energy-wise. Urban and rural contexts are quite different for the issue of locally generated energy. For example, wind energy generation is less efficient in an urban context, for technical reasons.

There are two sub-markets: new and retrofit, which differ substantially in their technical constraints. In these markets, the supply side (architects and builders, energy companies, heating, HVAC manufacturers, banks and real estate companies, etc.) have a major influence. In the commercial building sub-market, the demand side has more influence on construction and energy infrastructure, mainly through facility managers.

Most energy consumption comes from space and water heating and air conditioning, for which decision-making is mostly in the 'building market'; but a large part also comes from energy-using domestic appliances (washing machines, refrigerators, lights, computers, etc.), driven by rising standards of living and comfort, multiple purchases of appliances and the growing use of air conditioning, consumer electronics and new media. This is another market and regime where consumer behaviour is important and where design decisions are made by appliance manufacturers.

Finally, governments and policy-makers (as well as non-governmental organisations [NGOs] influencing policy) have a strong influence through the regulation of construction and through tax or subsidy policy.

2.1.1 Context meta-factors and specificities of the housing and energy domain

One big difference from other domains is that globalisation has little influence on housing. However, housing is obviously influenced and shaped by societal mega-trends, such as:

- Informatisation: pervasive computerisation will change life in homes with the arrival of home automation; it may be also one of the technical means of achieving greater demand-side control

● The ageing population: older inhabitants are likely to have less autonomy; an ageing population also means more homes with single or dual inhabitants, which means more energy consumed per capita

2.1.1.1 Population and housing: international and global mega-trends

The price of fossil energy (and of uranium) is bound to rise. Worldwide demand for energy is growing at about 2% per year as a result of rising population levels, globalisation, the growth of developing countries, notably China and India, and mechanisation. Consumption of marketed energy is forecast to expand from 495 quadrillion British thermal units (Btu) in 2007 to 543 quadrillion Btu in 2015 and then to 739 quadrillion Btu in 2035, a 49% increase over the 2007–35 period—an average of 1.75% per year (EIA 2010). The equivalent forecast average energy consumption growth rate for Europe is around 0.2% per year (EIA 2010)

Europe has gone through a demographic transition, but its population is still rising; the EU-25 population has risen by 6.8%, from 426 million in 1980 to 455 million in 2004, and is predicted to rise to 481 million in 2050 (+ 6%). Although it is difficult to provide simple statistics for the number of houses because of the difference of statistical systems within the EU, the number of houses is obviously growing. For example, in France, the total number of houses rose from 25 million in 1980 to 29.5 million in 2002 (+ 18%). In the same period the increase was 18% in Sweden, 19% in the UK and 39% in the Netherlands (NBHBP 2005). This will produce tension on the energy market. In Europe, the structure of the electricity grid cannot readily cope with strong growth, even if there were more generation plants. This means that distributed energy production, and local micro-generation, will be encouraged by authorities in many countries.

2.1.1.2 Climate change: international and global mega-trends

Climate changes are expected as a result of global warming. These will increase energy demand in buildings, both for heating and for cooling. These climatic changes will also put a strain on water resources, therefore causing problems for energy generation (for example in terms of power plant cooling, hydropower generation and storage), therefore raising energy prices and leading energy suppliers to sell energy at different prices at different times to avoid peak demand (such as a use of different day–night tariffs, seasonal tariffs, etc.). So climate change will produce significant tension in the HVAC area in homes (more demand, with greater variations in cost at different times); this should stimulate the market for innovation and sustainable solutions.

2.1.1.3 Political developments

Energy prices and housing policies are very sensitive political issues since they have a major impact on household budgets; therefore governments are cautious and reluctant to adopt unpopular measures in these domains.

The link between energy use and climate change is becoming a key political issue, with significant media interest in some countries; films, such as *An Inconvenient Truth*

(Guggenheim 2006), and global music events such as LiveEarth in 2007[3] are likely to keep the visibility of climate change high on the political agenda.

However, conservatism and inertia in the construction and building industry is important. This is being addressed in Germany and the UK by proposed energy efficiency standards that are more ambitious than current EU standards, such as EU Directive 2002/91/EC on the energy performance of buildings (European Parliament and European Council 2002). Proposed standards mean that buildings renovated in Germany will have to improve energy efficiency by up to 30%, and disclosure of heating and water costs have been required for all buildings since 2008. Also in 2008, the previously voluntary Code for Sustainable Homes (CoSH) (UK DCLG 2009) became mandatory in the UK and is expected to be extended to commercial buildings. CoSH Level 6 performance will require homes to be net positive energy generators.

From 2016 new houses in the UK are to be designed for zero net emission of CO_2 from all energy use in the home (UK DCLG 2006).

Although these moves seek to lead current practice, in general the lag in developing and agreeing building standards is likely to be a constraint on spreading good practice.

2.1.1.4 Social and cultural developments

The increase in number of households typically exceeds the rise in population, following increases in disposable income and the number of single-person households, particularly those containing the elderly and divorced.

Higher incomes mean that households can afford to purchase more appliances, although there is a mitigating trend as appliances become more efficient—with around 20% improvement from 1990 to 2003 (UK DTI 2006)—and as household energy efficiency as a whole improves. Overall, energy consumption per household in the UK has been almost static since 1990 (UK DTI 2006).

Stakeholders and end-users are becoming aware of climate change and therefore more prone to accept sustainability policies. The merits and methods of reducing individual 'carbon footprints' are commonly discussed in society and the media. However, energy consumption in housing is a strong component of comfort, and hence the introduction of more sustainable practices faces an obstacle combined with consumers' natural resistance to changing their habits. One generation is around 20 years, the EU birth rate is low and the average life-span in the EU-25 is now 72 years and increasing. Hence the natural turnover of population will be insufficient to renew habits, meaning that it appears to be necessary to convince present inhabitants of the need for change. On some occasions, though, such as at marriage, people are known to make significant changes to their habitat—the so-called 'windows of opportunity'.

Consumers, and therefore politicians, are reluctant to support the construction of new power generation facilities—plants, or even wind turbines—because of NIMBY-ism ('not in my backyard'). This reluctance also restricts changes in other parts of the infrastructure, such as new power lines.

---

3 liveearth.org/en/liveearth/070707

2.1.1.5 Economic and technical developments

The installed infrastructural base in construction and in energy supply is a significant factor in energy consumption in houses and buildings. Construction infrastructure has huge inertia: buildings often last more than 50 years. Annual new build constitutes about 1% of the housing stock. In France, this represents about 300,000 new houses every year. In the UK at least 75% of current properties are still expected to be in use in 2050 (UK BRAC 2006). At this pace, and taking refurbishment into account, the renewal of the present stock of housing built before 1975 (when the first directive on energy efficiency took place) will not take place before 2030. A slow renewal of stock also applies to the most energy consuming appliances. The lifetime of washing machines and dishwashers is about 12 years, and that of a tumble dryer about 14 years; fridges, cookers and ovens may last about 15 years, microwave ovens and televisions about 10 years, and a domestic personal computer (PC) about 6 years. Some of these appliances may see less use at the end of their life-cycle, their owners having already bought a newer, and hopefully more energy-efficient one; but they may also find their way onto the second-hand market.

Retrofit is costly, more than new construction; but the administrative regulations and the technical issues for destruction and reconstruction are often so complicated that retrofit is the solution chosen. So new buildings add to the existing stock rather than replace it; therefore most present buildings will still be around in 30 years. Most actors now focus on the refurbishment issue; but on technical grounds refurbishment is more difficult than new construction and needs a considerable amount of skilled work.

Energy generation and distribution systems are very large-scale investments with a long return on investment (typically decades) and have been heavily structured by present demand (for example to feed urban concentration). They are complex systems that have influence beyond a single EU member state. Pipelines or electricity grids have to be considered at an international level. End-user equipment, such as meters and wiring, last decades. This means that system changes, such as changing all present meters for 'intelligent meters', involve a long and costly process, which adds to inertia. For example, replacing the 30 million electric meters in France by 'intelligent meters' means 30 million interventions by skilled technicians.

Use of renewable sources and local generation of power are the obvious solution for sustainability, but there are many technical limitations:

- Renewable energy (wind, solar, biomass) is often cyclic and unevenly distributed. Needs, especially for heating and lighting, are cyclic, but are not necessarily matched to production capacity, which brings complex issues in relation to generation and network management. Research and positive examples are provided in case studies here, but the state of the art is far from satisfactory and requires investment, and fossil fuel is still easier and more flexible to manage, especially in an urban context

- Renewable and distributed energy such as wind and solar power need specific local studies by skilled operators. The current technicians and small and medium-sized enterprises (SMEs) are familiar with fossil fuel; therefore it has a competitive advantage beyond mere cost. Training of technicians will gradually take place as the renewable energy market grows

● Storage is a limiting factor and a major bottleneck in the development of 'renewables'. Present storage technology is both costly and inefficient. Electric batteries have a large ecological footprint, a high cost and only a small capacity. Other storage solutions (hydropneumatic, gravity, magnetic, electrostatic) involve large-scale investments and are mostly industrial in scale. Hydrogen generation and micro fuel cells are still in their early stages

Local renewable energy generation is still in the early phases of production, and progress is to be expected. In the medium term, there are technical limitations, especially in urban areas. It is reasonable to forecast that present large-scale generation plants will be used for many decades to compensate for natural fluctuations in wind and solar energy production.

Progress on the generation side and building technology will not be enough; all actors agree that demand-side control and less consumption is absolutely necessary. This will mean a need for changes in the behaviour of end-users. Some case studies in the following chapters show that consumer feedback on energy use is needed to obtain this goal.

2.1.1.6 Sustainability awareness and consumer will to change

Energy consumption depends not only on the energy efficiency of buildings and equipment but also on user behaviour. Frequency of washing, the setting of indoor temperature, etc. have a substantial impact. Barr *et al.* (2005) note that these actions are at two levels:

● Habitual actions, such as 'thermostat setting, closing off of unused rooms, altering room use, window closure when heating is on, using a clothes line rather than a tumble dryer, not filling the kettle full before boiling, putting a full load of washing rather than a half load' as well as ensuring proper maintenance of objects

● Purchasing activities, such as buying more energy-efficient equipment or installing energy-saving systems (double-glazing, etc.)

One could expect that, considering the rising awareness of consumers, which is apparent in most surveys, and the rising prices of energy, these types of behaviour will be more frequent if not systematic. Unfortunately, this is not the case and change is very slow. A vast series of studies, such as the BARENERGY EU research programme,[4] try to understand and overcome the barriers to consumer change and fill the so-called 'awareness gap' between declarations of interest and behaviour.

4 www.barenergy.eu/resources.html

2.1.2 Production–consumption chain and interlinked practices: the 'regime'

2.1.2.1 Physical affordability

The large majority of our building system is either small private individual habitat used for sheltering households in fixed places ('homes') or larger structures for collective activity ('commercial buildings': offices, shops, recreation space, etc.). Most of modern human activity takes place indoors, and, increasingly, in transport. Most buildings have a fixed envelope lasting several decades or centuries. Prefabricated, mobile and transient buildings represent an insignificant proportion of stock. Also, buildings are considered as a stable place for medium-term or long-term installations such as furniture, storage and domestic or non-domestic production equipment (kitchens, offices, etc.) designed to be operational on a continuous basis as a life and activity support system. This situation, which we take for granted, is in contrast to the nomadic or transient shelters used millennia ago and accounts for the inertia of this domain.

As houses are used to create a comfortable ambiance and service provision they are expected to come with heating, cooling, ventilation, hot water and various electric appliances. This would not be so problematic if houses and their internal systems were designed in a sustainable (or at least adaptable) way. Indeed, current buildings are built in such a way that the components of infrastructure cannot be easily disassembled or replaced without costly retrofit; even though it is known that these components (heating and cooling infrastructure, power, lighting, controls, water and other fluids, networks, indoor partitions, etc.) are at below optimal efficiency and have a much shorter lifetime compared with the building itself.

It is difficult and expensive to retrofit an old building to turn it into an energy-efficient one. Typically, construction of a new house takes 3–5 years from decision to build to occupants moving in; in the case of complex retrofit (e.g. including asbestos removal) the process may take up to 10 years to complete to a similar level of efficiency.

Also, opportunities to change buildings are usually limited to specific 'windows of opportunity' when people move or change the configuration of their household, often coinciding with life changes such as divorce, birth, etc., or when the building envelope or infrastructure needs major updating—every 20 to 40 years or so. The case is somewhat similar for commercial buildings, the refurbishment of which may follow events in the life of the organisation.

In homes, HVAC and hot sanitary water account for about 75% of energy use. In commercial buildings this proportion is lower (about 50%) because of the consumption of specific appliances such as computers. Lighting accounts for up to 25% of emissions, although these figures vary from country to country, depending on climate and energy efficiency. Table 2.1 shows the evolution of these figures in residential buildings between 1990 and 2004 in International Energy Agency (IEA) countries (mostly consumer economies). This situation derives from the cultural option of providing a comfortable constant thermal ambiance indoors, as opposed, for example, to thermal control through use of appropriate individual clothing. Electricity used for domestic appliances in households show the sharpest increase in the EU.

TABLE 2.1 Energy use in residential buildings in International Energy Agency countries: evolution from 1990 to 2004

Source: Laustsen 2008

	Energy use (%)	
Item	1990	2004
Space heating	59	54
Appliances	15	20
Water heating	18	17
Lighting	4	5
Cooking	4	4
Total	100	100

Most construction is urban, coupled with more people moving to urban areas. The density of houses in urban areas often excludes or limits some uses of renewable energy (such as wind power). Low-energy houses often rely on renewable energy that is subject to strong temporal variation (solar, wind), and therefore their relation to the energy grid—sometimes suppliers, sometimes consumers—may need a deep restructuring of the electricity grid. Therefore current infrastructure is often an obstacle to more sustainable energy provisioning for buildings.

A vast and complex set of rules, institutions, vested interests (including capital) and habits that are the result of historical 'evolution' and compromises between stakeholders make every change a complex negotiation process, where each player has limited power to act. Anticipating a three-layer model for change to be discussed in the conclusion (Chapter 11), beyond the issues surrounding the physical affordability discussed here, two further aspects of the regime are relevant; people's representations and practice, and institutions (rules).

2.1.2.2 People's representations and practice

Historically, stakeholders have not viewed houses and buildings as representing an energy-using product, appraising the potential for improving energy performance as a primary concern or positive asset. As Laustsen (2008) summarises:

> Many barriers hamper energy efficiency in new and existing buildings. When new buildings are designed and constructed, energy efficiency is but one concern among many factors in construction. Energy efficiency in buildings may be low on the list of requirements for the building. The development of most buildings focuses on construction costs with very little concern for running costs. Different people and budgets may govern the operation of a building, often entailing split incentives for energy conservation. Very rarely will any single decision-maker participate in all aspects of a building's construction, operation and financing. Most decision-makers will not have the data or capability to calculate a building's lifetime costs and estimate the consequences of early design decisions. Consumer inertia regarding build-

ings' energy performance stems from the fact that energy is invisible, that the energy costs of new buildings seem imaginary and that improved efficiency can decrease prestige.

Through the meshing of the building structure with its internal infrastructure, most changes in the building itself (and especially the refurbishments necessary for better energy efficiency) may have implications for individuals' behaviour and practices, spreading out to other domains. Therefore, users are reluctant to make modifications because this could have a significant impact on their lives. Changing something in a house may feel like trying to mend the hull of a boat at sea.

As an example of limitations to changing practices, there is yet little or no specific metering of domestic energy use. A consumer will get a monthly bill but will not be able to relate that to specific uses such as the energy used for baths, a washing machine cycle, or for the television and 'set-top box' stand-by consumption. Although there are a few niche providers of energy meters, the existing systems and practice do not identify the energy-hungry uses and therefore do not highlight areas of overperformance or under-performance and thus do not highlight areas for improvement and where a change in practice is needed.

2.1.2.3 Regulatory institutions

Public authorities have made two significant moves: they have created more incentives and regulation towards sustainability, for example through the EU Emission Trading Scheme and through deregulation of the energy market. Regulation by market forces has a twofold impact in response to incentives to consume and the pressure for growth:

● Less government control limits large-scale coordination and regulation through mandatory policies

● Easier development of local initiatives allows external funding sources to be found and opens up administrative paths to help initiators in installing new systems

The Energy Performance in Buildings (EPB) Directive was introduced for implementation in EU Member States by 2006 to stimulate market demand through standardised energy rating and certification against energy performance requirements for new buildings and major renovations (European Parliament and European Council 2002). The elements of the EPB Directive are illustrated in Figure 2.1, which shows the scope of the system and the reference to existing EU norms.

The EPB Directive (European Commission 2003) offers

● A common EU-wide methodology for calculating the energy performance of a building, taking account of local climatic conditions

● Minimum standards for energy performance, determined by member states, and applied both to new buildings and to major refurbishments of existing large buildings, many based on existing or planned European norms

FIGURE 2.1   European norms associated with the Energy Performance in Buildings Directive articles

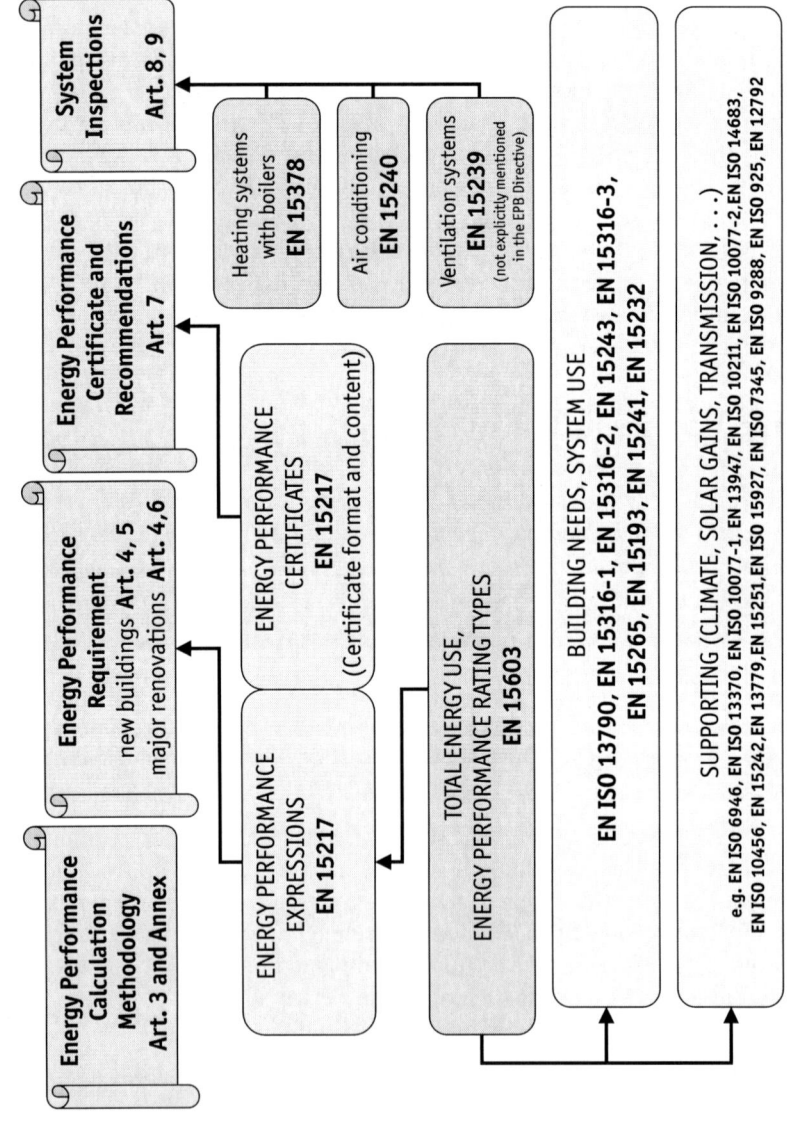

Source: Hogeling 2010

- A system of building certification to make energy consumption levels much more visible to owners, tenants and users

- Requirements regularly to inspect boilers and air-conditioning systems above minimum sizes to verify their energy efficiency and greenhouse gas emissions

2.1.2.4 Market institutions

The market is the main mode of 'regulation' in this domain, which is both a favourable ground for local innovation and an obstacle driving radical changes to long-term public policies. Construction companies and the network of stakeholders in new building (architects, developers, banks, HVAC sellers and installers, insurance companies, etc.) are already well organised to improve new construction, so that projects are easy to plan within existing routines and regulations. Although still seen as a niche, some developers and funders are now becoming interested in 'greener' properties because of the market premium such properties command.

However, refurbishment and renovation of existing buildings is done by a complex ecology of SMEs, many of which do not have the know-how for sustainable retrofit. In fact, these SMEs often install 'counter-references' (that is, inefficient installations) because of their lack of experience with new systems such as solar generation. In addition, awareness of green options may be limited among those individuals who are increasingly completing DIY (do-it-yourself) installations affecting energy use in various countries in Europe. Every retrofit case is different and exact tuning and details are crucial to get a positive result (Caird *et al.* 2008). Solutions and market institutions valid under a specific climate may be counter-effective elsewhere. Expertise is slow to build and not easy to scale up.

Social housing organisations, 'leaseholders', property agents (including domestic landlords) and commercial facility managers and DIY or SME suppliers (builders, retailers and manufacturers) are also important actors, the latter particularly for refurbishment. Under current energy prices, the return-on-investment horizon for a good retrofit may discourage such investment in rented houses. Landlords do not support the running costs of inefficiency, and renters may not get a return on their investment within their rental period.

Germany and California have used subsidies to reduce the price of solar energy, to increase adoption and to pull through technologies, improving the economies of scale.

The electricity distribution infrastructure and market is still a crucial element in the system, and most buildings rely on it for their consumption. Most energy companies are the product of the 'unbundling' and fusion of 'historic operators'. They are large and have vested interests and sunk costs in generation plants and networks. Distributed generation is still hardly profitable at present oil prices without public subvention for these large entities. It also has negative externalities in network management, because it makes the provision of demand on the grid less predictable. For example, a whole area may draw or withdraw demand quickly according to wind speed. Energy companies are, as organisations, naturally risk averse and slow to take initiatives to establish a new system that questions their present practice, market position and investments.

2.2 Systemic sustainability problems and windows of opportunity

2.2.1 Key systemic sustainability problems

The three aspects of sustainability—environmental, social and economic—are all to be addressed with respect to energy use in the housing domain. The environmental 'Factor X' improvement is well known but is mostly to be addressed by the supply side in response to depleting non-renewable fossil fuels and the associated production of greenhouse gases emissions, linked to climate change. Consumers can act by 'voting with their feet' (Hirschman 1970) or imposing political pressure as citizens. The German case, described by Reusswig, Lorek and Fuchs in Chapter 10 of this book, is quite instructive in this respect.

Climate change, and especially the increase in seasonal variation of temperature, will make energy demand even more salient as many buildings that could maintain a comfortable thermal ambiance in previous conditions may now need modifications to stay comfortable enough in hotter summers or colder winters. For example, the 2003 heatwave induced many households in France to buy air conditioning, inducing a rise of 40% in the (small) market of fixed air-conditioning units (a catastrophe because air-conditioning units use greenhouse-effect gases). In individual houses, operable windows, ventilation, shading and other alternatives can be used, but many office buildings are designed in such a way that increased air conditioning is presently the only solution for cooling (Maldonado and Oliveira-Fernandes 1993). If buildings follow the same trend as cars towards systematic air conditioning, the ecological impact may be disastrous.

The social factor, 'equitable growth' in the North and South, is relevant both on a macro level (for example, the conflict issues concerning energy in the Middle East and Russia) and on a meso level. As in many domains, sustainability solutions are presently for the richer communities, because they need some investment. Inhabitants of low-rent flats or apartments have little control over their heating system. This is more a matter of policy. Case studies show that this policy may be quite local.

The major issues in buildings seem to be economic in nature. The problem is to reduce energy consumption and still keep an acceptable level of comfort at an acceptable level of financial investment. Reduced demand goes against the present habits and trends of the European population. The cases of power outages or shortages show how dependant we have become on energy consumption for our daily life and that people experience significant difficulties in managing their lives with a reduced energy supply. The availability of cheap energy enabled the industrialisation of society and a level of automation that might be seen as addictive. This is certainly an area where education and strong progress have to be made.

Finally, the significant price of buildings, primarily based on the price of land rather than on the price of construction itself, weighs down any decision to change with its economic impact.

2.2.2 Stability and windows of opportunity in the system

There are some stable trends, forcing producers and consumers in a specific direction. We will call them 'stabilising factors'. In contrast, 'destabilising factors' play out as tension between actors, tension between generally accepted norms or values and practices and so on.

2.2.2.1 Stabilising factors

Rising oil prices and their anticipation guarantee that stakeholders will continue to invest in sustainability. It may seem a paradox, but the shared anticipation of the end of the present regime based on affordable oil is a good incentive for investors.

Inertia factors in the energy and construction frameworks and infrastructure act both ways. It is a technical obstacle to more sustainability, but the rigidity of the building stock also stands as a large potential market for better solutions. This is well documented and provides a solid basis for business plans. Our daily activity, especially in an urban environment (which is the main arena in consumer economies), is subserved by urbanism. The fixed nature of building organisations and of their link to infrastructure (roads, electricity, telecommunications, water supply, waste systems, etc.) is, then, a factor contributing to the inertia of many other domains such as mobility.

Consumer demand in domestic energy use in terms of comfort is clear and stable (ambiance, temperature, hygiene). This enables the supply side to construct 'offers'.

2.2.2.2 Destabilising factors

Global warming is accelerating. It is a problem because the climate becomes more erratic, which is a technical problem for thermal regulation. It also increases the demand for air conditioning, which is energy-greedy and has a large ecological footprint.

Deregulation of the electricity market acts both ways. On the one hand, it stimulates competition between suppliers and opens new niches for renewable generation. On the other hand, it shortens the anticipation horizon of suppliers, which may not favour long-term investments in renewable energy; it also makes the landscape more complex and prices more volatile, which means less visibility and more risks for investors in 'green' technologies.

Technical progress (such as in materials, controls, 'green' roofs and in the use of biomass) is a positive destabilising factor. It is a continuous source of innovation in building products, also making the decision to change more complex and may discourage or delay local investment decisions. Another problem it raises is the installation of 'counter-references'—where inappropriate technology, or the inadequate installation of new technology, brings problems to the end-users and casts doubt and a bad reputation on some sustainable technologies. Home automation is one of the key issues here. It may bring huge progress in consumer feedback and energy-use control, but most actors in the domain are still hesitating about strategies because the technology is not fully mature, and there are no 'killer services' yet.

There is a gap and hence tension between what needs to be done and what is actually feasible right now, for technical, economic or social reasons. Three nodes of tension are to be monitored:

- The 'pace of progress' between energy suppliers (and, more generally, builders) and sustainability champions

- The action–awareness gap between what consumers and other actors recognise they *should* do (values) and what they *actually* do (practice)

- The market tension between local experiments and the wider competitive market

The 'pace of progress' gap is a tension arising partly from misunderstanding about the possible pace of replacement of the present energy generation technology by 'renewables'. Energy companies are bound both by the economic rationale of financial balance and by technical issues. For example, distributed generation brings a series of new problems of managing the electricity grid with renewable energy generating installations (such as wind turbines and solar generators), which have a high degree of variability in grid input and quality. System-wide regulation is needed to manage a grid, because the amount of energy fed into the grid must equal the amount used, every second. Present grid management is through a sophisticated system of demand forecast that commands power generation. The grid is an international system, and local failure or disequilibrium when one power input fails to feed in the grid as planned, in terms of both time and volume, can instantly disrupt equipment and transformer stations, propagate to large areas of the grid and provoke a massive blackout in a matter of minutes. Inserting many new small producers into the system makes it much more complicated to manage. At present, most local generation systems still rely on the grid. The transition from a situation where local generation is marginal to one where it would represent a substantial amount of the consumption also means a deep change in the way the grid is managed; and the new system will need a very efficient information and provision system to operate because of the physics of the system that do not tolerate approximation or time delay.

In fact, the present state of the art in grid management and information systems is not yet capable of handling widely distributed power generation. A great amount of research and development (R&D) and public investment is necessary to create an 'intelligent grid' that could manage the new system, investment which is often underestimated by advocates of renewable energy who believe that the cautious attitude of energy suppliers is merely driven by the desire of selling more centrally generated energy. Failure to understand this gap results in an opposition between energy suppliers and the demand for sustainable and local generation. However, these suppliers are both an obliged player and an enormous potential investor and market driver in the domain if a viable economic path can be found for them to insert the management of renewable and local generation in their grid framework.

The action–awareness gap is between what consumers would be like to do for sustainability and what they actually do on an everyday basis. This gap is well documented, and the obstacles are known. Some paths of improvement may result from the provision of detailed feedback on consumption, as shown in the cases presented here, especially those examined by Fischer (Chapter 9), and by making available a wider set of offers through a network of skilled salespeople and installers. The growing awareness of climate change is pushing change in the right direction. The conclusion (Chapter 11), based on the SCORE! discussions, proposes some ways of solving this issue.

Finally, the market tension between local experiments and the wider competitive market is a crucial issue. Actors must know the rules of the game to implement solutions together in a systematic way. One can distinguish at least three different scenarios depending on the stage of development of a sustainable building solution, from local experiment, through niche markets, to large competitive market. There is a tension at the threshold between each of these stages, where failure may occur. This is expanded on in Chapter 11, in which the potential for change through a three-step model for transition is considered. Some promising approaches are introduced below.

2.2.3 Promising niches and actors that could stimulate change

Systemic approaches remain marginal. The most common current and proposed approaches to making the consumption of energy in buildings more sustainable are:

- To use less energy at the consumer level: turn down or switch off

- To switch to renewable sources at the consumer level: buy greener

- To design buildings that are more sustainable

- To install energy-efficiency control and efficient appliances

2.2.3.1 Consuming less

Drivers for people to change behaviour towards more sustainable energy use in houses and buildings by consuming less include:

- An increased price for energy

- Increased awareness, for example through media interest, reflecting the potential emergence of a third 'green' consumer wave focused on climate change—at least in certain countries

- Improved energy-monitoring technology

Most arguments for consuming less have, until now, been based on the economic costs and benefits, but monetary incentives are not always sufficient. Social approaches, for example participating in a local community involved in more sustainable behaviour, could bring enough immediate benefits in shared social welfare to motivate participants. Involvement in a community is a crucial factor in behavioural change, as well documented by studies in social psychology (Lahlou 2006; Lewin 1952). This social benefit can compensate for less comfort and heavier constraints coming with a less energy consumption.

2.2.3.2 Switching to greener energy

The increasing availability of more sustainable energy generation and options for consumption, such as 'green' electricity tariffs, are enabling people to change behaviour towards more sustainable energy use in houses and buildings. The economic case, the available choice and the facilitation of the process for switching to energy generated from cleaner and more renewable sources are improving—typically, this relates to electricity from renewable sources, although biofuels for heating are finding increasing use. Examples of self-sufficiency in domestic renewable energy generation remain very limited, and the handling of distributed generation remains problematic without economic storage technologies. This also puts positive pressure on the energy providers.

2.2.3.3 Design of eco-houses, housing developments, towns and cities

According to the European Commission (2003), 'research shows that more than one-fifth of the present energy consumption and up to 30–45 MT of CO_2/year could be saved by 2010 by applying more ambitious standards to new and when refurbishing buildings—which represents a considerable contribution to meeting the Kyoto targets'. Raising the awareness of architects and other stakeholders in the design phase of the construction system represented by major house builders is key. If they do not change their 'products' then there is the problem of 'locking' in suboptimal energy-using practices.

Regulatory standards such as those implemented by EU member states under the EPB Directive (see Section 2.1.2.3) are stimulating market demand through standardised energy rating and certification against energy performance requirements for new buildings and major renovations.

In the UK, demonstration homes are being built to show how Code for Sustainable Homes Level 6 performance can be achieved, incorporating net positive energy generation. There should be a recognition that people already want to see 'greener performance' adopted in designs for homes that are mainstream and attractive, not niche. Also, the UK proposal to build at least four new eco-towns will establish a system level approach on a larger scale, already under way on an even larger scale in planning the Dongtan eco-city in China (WBCSD 2006).

System-level approaches enable the use of 'energy cascading'. The production or consumption of energy may provide a further source of energy, lower in thermodynamic quality, for another 'lower' use. For example, electricity can be put to many uses (such as to create force, provide lighting, and so on) and will produce heat as a by-product during use, which can be used only for heating. Energy recycling in this cascade is carried out through centralised production—combined heat and power (CHP) plants—but rarely implemented at end-use. The 'Building as Power Plant' (Hartkopf 2001, 2004), with its ascending–descending energy cascade, is a landmark project in this domain.

Energy storage has been a limiting factor and a major bottleneck in the development of renewable energy. The development of efficient and economic storage technology would overcome some of the significant shortcomings of renewable generation. Hydrogen generation and micro-fuel-cell prototypes are being developed, as are domestic CHP units. Progress is to be expected in this area.

New build should not be seen as the only solution to the problem; rather, the answer should be sought in retrofitting the existing building stock and changing the behaviour of consumers in inherently limited buildings—through awareness, education and making technologies available and affordable.

2.2.3.4 Installed technology; energy-efficiency control and efficient appliances

There is a need to design technology into existing homes to enable it to have a neutral or positive effect on impact-reducing consumer behaviour; technology that makes more sustainable lifestyles easier. One strand of development is towards 'intelligent buildings', exploiting information technology developments such as home and office automation. With machine-to-machine (M2M) communication, buildings will become actors in their own adaptation. This introduces a new 'actor' in the system. Whereas the consumer may be careless, the building can have some sustainable 'reflexes'; it can act to open windows at night for cooling, switch off lights when the building is empty, store energy and switch to renewable sources and optimise thermodynamic transitions.

Information technology will enable 'cradle-to-cradle' design by whole-life monitoring of products and systems, informing their redesign and management. It will also feed back individual decisions and consumption. For example, the consumer could ask her or his house to 'give me a €2 shower'.

Better-designed monitoring and control devices, for example thermostats and dials on washing machines (for example, promoting washing at 30°C), may start to influence consumer behaviour. This relates to all domestic energy-using subsystems, including consumer electronics, information and communication technology, home appliances, etc., given most have their greatest environmental impacts through the energy consumed in use. Use of a direct current home network for brown appliances would save the power loss in transformers and 'standby' electricity consumption. At this micro level appliance energy efficiency is being addressed through the Energy-Using-Products (EuP) Directive and other initiatives to reduce standby consumption, such as the IEA global initiative 'One Watt Plan' (IEA 2002).

Changes to behaviour and the adoption of such new technology is likely to occur only incrementally. Not all appliances are changed at the same moment in time. This is an obstacle to a more systemic transformation in household energy consumption.

2.3 Conclusions

There is likely to be ongoing concern over climate change in a number of countries as a result of increased media exposure, a growing financial interest in sustainable energy technologies and so on. Buildings have been implicated as a major area of energy consumption, and concerted action is needed to improve the position.

In several countries new initiatives are starting that focus on low-carbon homes. There are production as well as consumption issues. One stake on the production side is to set up a 'green' building supply chain. It is much easier 'to green' new housing developments if: (a) new technologies with robust supply chains are in place, (b) there is a supportive environment, (c) stakeholders are engaged and (d) each partner knows what to do. For example, if builders' merchants provide eco-products and materials, architects specify them and builders buy them.

As new construction is marginal in EU economies (compared, for example, with China), there is no rapid growth of infrastructure that could easily change the regime by outnumbering the present stock of real estate with new energy-efficient buildings. Therefore, changes take place mostly through the modification of existing infrastructure (through retrofits, adapting the structure of energy provision, changing stable habits, etc.) by using market mechanisms and trying to orientate them with state regulation. However, retrofitting of buildings is costly and, each case being specific, there is little economy of scale.

It is not enough to act on the building construction chain. Effort is needed on the user side, promoting the adoption of more energy-conscious behaviour on a day-to-day basis. Even though consumer awareness is growing, there is still a large awareness–action gap. Enabling householders to reduce energy use in the home, whether it relates to heating or cooling or the use of consumer electronics and household appliances, will be on most governments' agendas; but persuasion is not enough, and it is as yet unclear what measures are most efficient in fostering changes towards more widespread sustainable behaviour.

Innovation in market economies relies on individual initiative, but it is unclear in many cases what benefits most stakeholders with power in the domain would receive from more sustainable energy use in buildings. Clearly, end-users would get a lower energy bill; but most actors and decision-makers in the domain have little immediate incentive to start on an uphill green path towards sustainability, pushing against the domain's rigidity and inertia.

The case studies in this book will show that, even though the global picture seems rather dark at first sight, another world is still possible. We have gathered in this book a few of the many cases of change towards SCP. In some cases, a reasonable forecast would have been failure. In any case, the motivated champions behind these cases seem to have applied the famous motto of Netherlands governor William I of Orange-Nassau (William the Silent, 1533–1584): '*Point n'est besoin d'espérer pour entreprendre, ni de réussir pour persévérer*'.[5] And, indeed, the cases were finally successful, so we can learn from them.

Chapter 11 sums up some of the lessons learned and provides a framework for strategies to help stakeholders in their enterprise. Now we encourage the reader to look at the case studies; no general analysis will replace the flavour of real stories.

---

5 'Need not expect to undertake; need not succeed to persevere'.

References

Barr Stewart, G., W. Andrew and N. Ford (2005) 'The Household Energy Gap: Examining the Divide between Habitual and Purchase-Related Conservation Behaviours', *Energy Policy* 33: 1,425-44.

Caird, S., R. Roy and H. Herring (2008) 'Improving the Energy Performance of UK Households: Results from Surveys of Consumer Adoption and Use of Low- and Zero- carbon Technologies', *Energy Efficiency* 1.2 (May 2008): 149-66.

Commissariat Général au Développement Durable (2009) 'Repères. Chiffres clés de l'énergie. Service des Statistiques', December 2009; www.developpement-durable.gouv.fr/IMG/pdf/Repere.pdf (accessed 3 November 2010).

EIA (Energy Information Administration) (2010) 'International Energy Outlook 2010', Office of Integrated Analysis and Forecasting, US Department of Energy, Washington, DC, available at eia.doe. gov/oiaf/ieo/index.html (accessed September 2010)

European Commission (2003) 'Better Buildings: New European Legislation to Save Energy'; ec.europa. eu/energy/demand/legislation/doc/leaflet_better_buildings_en.pdf (accessed October 2009).

European Parliament and European Council (2002) 'Directive 2002/91/EC of the European Parliament and of the Council of 16 December 2002 on the Energy Performance of Buildings'; eur-lex.europa. eu/LexUriServ/LexUriServ.do?uri=CELEX:32002L0091:EN:NOT (accessed October 2009).

Guggenheim, D. (2006) *An Inconvenient Truth*, available from www.climatecrisis.net (accessed October 2009).

Harris, E. (2008) 'Upgrading the Grid', *Nature* 454 (31 July 2008): 570-73.

Hartkopf, V. (2001) 'The Building as Power Plant', Concept Paper, Centre for Building Performance and Diagnostics, Carnegie-Mellon University, Pittsburgh, PA; www.arc.cmu.edu/bapp/Overview/ Concept-Paper.pdf (accessed October 2009).

—— (2004) 'Building as Power-Plant (BAPP)', *Co-generation and Distributed Generation Journal* 19.2 (Spring 2004): 60-73.

Hirschman, A.O. (1970) *Exit, Voice, and Loyalty: Responses to Decline in Firms, Organisations, and States* (Cambridge, MA: Harvard University Press).

Hogeling, J. (2010) 'Experience on the Implementation of the EPBD and EPBD-CEN Standards in the EU Member States and the Netherlands', CENSE, 23 March 2010; ftp://ftp.cenorm.be/CEN/Sectors/ List/Energy/energymanagement/EUSEW2010/13HOGELING.pdf (accessed September 2010).

IEA (International Energy Agency) (2002) 'Reducing Standby Power Waste to Less than 1 Watt: A Relevant Global Strategy that Delivers' (Paris: IEA; www.iea.org/papers/2002/globe02.pdf [accessed September 2010]).

—— (2005) *Key World Energy Statistics 2005* (Paris: IEA).

Lahlou, S. (2006) 'Les systèmes et niveaux de détermination du comportement alimentaire', *Cahiers de Nutrition et Diététique* 5: 273-78.

Laustsen, J. (2008) 'Energy Efficiency Requirements in Building Codes: Energy Efficiency Policies for New Buildings' (information paper; Washington, DC: Energy Information Administration; Paris: OECD , March 2008).

Lewin, K. (1952) 'Group Decision and Social Change', in G.E. Swanson, T.M. Newcomb and E.L. Hartley (eds.), *Readings in Social Psychology* (New York: Holt).

Maldonado, E., and E. Oliveira Fernandes (1993) 'Building Thermal Regulations: Why has Summer been Forgotten?', in H.S. Stephens (ed.), *Solar Energy in Architecture and Urban Planning* (Bedford, UK: H.S. Stephens Publishers): 626- 30.

NBHBP (National Board of Housing, Building and Planning) (2005) 'Housing Statistics in the European Union, 2004' (Karlskrona, Sweden: Boverket; www.iut.nu/EU/HousingStatistics2004.pdf [accessed October 2009]).

Noeren, D. (2007) 'CO$_2$ Emission Reduction in the German Household Sector till 2050: Barriers and Incentives' (thesis, Lund University International Master's Programme in Environmental Studies and Sustainability Science, 2007; www.lumes.lu.se/database/alumni/05.07/thesis/Dominik_ Noeren.pdf [accessed 3 November 2010]).

TAREB (2002) 'The European Directive on the Energy Performance of Buildings', in *Learn Package Low Energy Architecture: Energy Assessment* (www.learn.londonmet.ac.uk/packages/tareb/docs/lea/ lea_ch3_en.pdf [accessed October 2009]): chapter 3.

Tukker, A., G. Huppes, S. Suh, R. Heijungs, J. Guinee, A. de Koning, T. Geerken, B. Jansen, M. Van Holderbeke and P. Nielsen (2006) 'Environmental Impact of Products (EIPRO): Analysis of the Life-cycle Environmental Impacts Related to the Final Consumption of the EU-25' (Main Report; European Science and Technology Observatory/Institute for Prospective Technological Studies [ESTO/ IPTS]; ec.europa.eu/environment/ipp/pdf/eipro_report.pdf [accessed October 2009]).

UK BRAC (Building Regulations Advisory Committee) (2006) SDC Report on Existing Building Stock (London: ODPM; www.communities.gov.uk/documents/planningandbuilding/pdf/151819.pdf [accessed September 2010]).

UK DCLG (UK Department for Communities and Local Government) (2006) *Building a Greener Future: Towards Zero Carbon Development* (Wetherby, UK: Communities and Local Government Publications; www.communities.gov.uk/documents/planningandbuilding/pdf/153125.pdf [accessed September 2010]).

—— (2009) *Code for Sustainable Homes Technical Guide* (London: Communities and Local Government Publications; available at www.planningportal.gov.uk/uploads/code_for_sustainable_homes_ techguide.pdf [accessed September 2010]).

UK DTI (UK Department of Trade and Industry) (2006) 'Annex 3A: How Lifestyles Affect Energy Demand', in *Energy: Its Impact on the Environment and Society* (London: DTI; webarchive.nation-alarchives.gov.uk/tna/+/http://www.dti.gov.uk/files/file20327.pdf [accessed September 2010]).

Van Raaij, W.F., and T.M.M. Verhallen (1983) 'A Behavioural Model of Residential Energy Use', *Journal of Economic Psychology* 3: 39-63.

WBCSD (World Business Council for Sustainable Development) (2006) 'Dongtan: The World's First Eco-city' (Geneva: WBCSD; www.wbcsd.org/Plugins/DocSearch/details.asp?DocTypeId=251&O bjectId=MTk4MTk [accessed September 2010]).

3
An innovative approach to designing zero-energy residential buildings in Boston: enhancing and monitoring learning[1]

Halina Szejnwald Brown
Clark University, USA

Philip J. Vergragt
Tellus Institute, USA; Clark University, USA

Over the past decade, environmentally oriented innovations in technology and services have emerged in all areas of the economy, driven by governmental policies, professional experts, market opportunities and social movements. The building construction sector, where interest in high-performance buildings has been on the rise, is a primary example. In a different realm, the so-called 'new urbanism' and 'sustainable communities' movements magnify the above trends. Although often driven by broader considerations, such as quality of life, economic development and equity, these movements converge with the current developments toward high-performance buildings by promoting high-quality construction that minimises indoor drafts and air pollution as well as other features that collectively reduce energy intensity (Hallsmith 2003; Portney 2003; Register 2002).

In the language of evolutionary economics, one might take the above trends as evidence of an ongoing transition toward a more sustainable sociotechnical system of

---

1 We thank the members of the Old Distillery Core Team for their willingness to be interviewed and for allowing us to follow their project meetings, and for their interest in this project.

building design, construction and maintenance. The concept of sociotechnical system denotes a relatively stable configuration of techniques and artefacts—as well as institutions, rules, practices and networks—that determine the 'normal' developments and use of technologies in a particular area of human needs (Kemp *et al.* 2005). Sociotechnical systems fulfil socially valued functions that they, in turn, constitute. They also embody strongly held convictions and interests concerning particular technological practices and lifestyles, existing institutions and the best ways in which these may be improved. Stability and resilience are central to sociotechnical systems. That means that change is slow, involving not only innovations in science and technology but also changes in institutions, professional norms and practices, lifestyles, belief systems and others.

Society thus faces a dilemma: the dominant sociotechnical system of building design and construction (as well as others, such as transportation) naturally resists the urgently needed rapid societal transition towards more sustainable ways to satisfy human needs and wants. Resolving this dilemma has kept analysts and policy-makers active during the past years. All agree that one of the conditions for affecting rapid change is that the professions and other communities of practice linked to building design, construction and maintenance fundamentally reconsider some of their norms, practices and problem definitions. Stated differently, higher-order learning on a scale ranging from individuals to professional and business communities, to society at large. is necessary.

One way to facilitate learning toward sociotechnical system change is through small-scale experiments aimed at developing, testing and introducing new technologies and services. We previously referred to this type of experimentation as bounded sociotechnical experiments (BSTEs; Brown *et al.* 2003). Numerous authors refer to the importance of higher-order learning in sociotechnical experiments and often note its absence (Hoogma *et al.* 2002, 2005). Yet, with a few exceptions (Hoogma 2001; Brown *et al.* 2003), little systematic study has been done on defining the learning processes in experiments, monitoring them, assessing their societal impacts or examining the conditions under which learning occurs (or not) and by what mechanisms. Gaining a better understanding of the learning processes occurring within BSTEs and, beyond them, through diffusion, is the subject of the research described in this chapter.

In this chapter we examine the learning processes in the design of a building with zero greenhouse-gas emissions, located in Boston, MA. The empirical observations are analysed through the lens of a conceptual framework that we propose in the next section. The framework draws on two types of sources: the theoretical and empirical literature on learning by individuals, organisations, communities of practice and societal actors engaged in policy debates, and the work of Grin and Van de Graaf (1996a, 1996b), who studied the learning process in a discourse over wind energy in Denmark. We conclude that higher learning takes place on two levels: on the level of individual heterogeneous actors, and on the level of the project team. We map the learning processes and offer some evidence for the likely diffusion of this learning into various communities of practice.

3.1 Conceptualising higher-order learning

Higher-order learning is a radical change in interpreting observations (interpretive frames) and in solving problems and advancing objectives. The term 'higher-order' denotes what in organisational sciences has been dubbed 'double-loop' learning (Argyris 1977; Argyris and Schön 1978) or 'generative' learning (Senge 1990) and in policy sciences as 'conceptual' learning (Glasbergen 1996). It entails changes in the assumptions, norms and interpretive frames that govern the decision-making process and actions of individuals, communities and organisations or that underlie a policy discourse. It occurs through reflection and self-evaluation. Higher-order/double-loop/generative/conceptual learning contrasts to lower-order/single-loop/adaptive/technical learning, respectively, in which problems are corrected or policies altered without changes in problem definition, interpretive frames or in norms and values.

Learning occurs through a feedback-stimulus mechanism, when the existing, well-accepted, time-tested and trusted interpretive frames and competences receive feedback on their performance in solving a problem or advancing specific objectives. If, as a result of this feedback, it becomes apparent that the desired results are not forthcoming, these cognitive constructs become subject to reassessment and, if necessary, are replaced with new ones. This broad concept of feedback-stimulus is consistent across a wide range of disciplinary writings about learning, from cognitive sciences to organisational sciences to policy sciences. Working within the context of cognitive sciences on how individual professionals learn through problem-solving, Schön (1983) showed in a seminal study that the process starts with an intuitive defining of the problem within the context of the interpretive frame typical for that professional group. The actual problem-solving consists of iterative 'conversations' between a professional and the problem, through trial and error, which in turn leads to increasingly higher-order reassessments: it is via these increasingly higher-order reassessments that learning occurs.

In the context of organisations, the stimuli necessary for higher-order learning come from threats to organisational survival and success, failures, disasters and other surprises (Argyris 1977; Argyris and Schön 1978; Sitkin 1992). Senge (1990) additionally writes about various group techniques that generate feedback on accepted assumptions and behaviour as the means to stimulate higher-order learning in organisations (see also the review by Easterby-Smith 1997). Like Senge and others in the context of organisational learning, Berkhout emphasises collective visioning and scenario-building exercises as a vehicle for inducing learning on a societal scale (Berkhout *et al.* 2002).

Wenger (1998, 2000) uses the 'community of practice' as a unit of analysis in order to examine the mechanisms by which external stimuli induce learning in groups and organisations. In Wenger's language, the feedback process that is central to learning takes place by way of interaction between the deep competency possessed by a community of practice and the experience it acquires by interacting with the outside world. It is these boundary processes that produce learning. In policy sciences, higher-order learning is broadly understood as a collective change in prevalent views, norms, problem definition and relationships among groups. Like organisational and cognitive sciences, this school of thought attributes learning to the presence of feedback loops between the existing belief system and interpretive frames, and new experiences. New knowledge and interactions among groups with different belief systems and interpretive frames play a key role in inducing learning (Keohane and Nye 1989; Lee 1993; Sabatier 1999; Schön and Rein 1994; Van Eijndhoven *et al.* 2001; Wildawski 1990). There is a widespread agreement that crises, a sense of urgency and the availability of platforms for interaction are important facilitators of social learning (Birkland 1997; Schön and Rein 1994). Paquet (1999) advocates social experimentation as an effective inducer of the processes leading to social learning.

Schön and Rein (1994) and Fischer (1995) see higher-order learning, and the subsequent reframing of a problem, as the answer to solving intractable policy controversies. Such learning occurs through a multilevel discourse among the stakeholders. Fischer identifies four levels of increasingly higher-order discourse: technical, on the level of specific tools, costs and benefits; contextual, on the level of problem definition within a given interpretive frame; systemic, on the levels of setting goals and objectives in relation to societal needs; and ideological, on the level of fundamental beliefs about the social order.

In the present study we draw on the concept of learning as a result of feedback-stimulus through interaction of different interpretive frames and problem definitions to map out the learning processes among the immediate participants in sociotechnical experiments. The next section takes a closer look at a BSTE as a place for higher-order learning.

3.2 Bounded socio-technical experiments as agents for social learning

Previously we introduced the concept of a BSTE to denote a project exhibiting several characteristics. It is an attempt to introduce new technology or service on a scale bounded in space and time. The time dimension is measured in years (not decades or months), while the space dimension is defined either geographically (a community) or by a number of users (small). BSTE is a collective endeavour, carried out by a coalition of diverse participants, including business, government, technical experts, educational and research institutions, non-governmental organisations (NGOs) and others. There is a cognitive component to BSTE in that at least some of the participants, and definitively the analyst, explicitly recognise the effort to be an *experiment*, in which learning by doing, trying out new strategies and new technological solutions and continuous course correction are standard features (Brown *et al.* 2003). BSTE participants view themselves as engaged in an experiment 'that could become the first of a series of the same'.

A BSTE is driven by a long-term and large-scale vision (in this case, of advancing society's sustainability), though the vision need not be equally shared by its participants. Its goal is to try out innovative approaches for solving larger societal problems of unsustainable technologies and services. This latter characteristic distinguishes a BSTE from, for example, solving a particular environmental problem in a community (such as alleviating pollution through traffic control) or from a strictly market-driven introduction of new technologies and services (for example, introduction of alternative electric-powered vehicles, such as Gismo, Sparrow and many others; Brown and Carbone 2006). Small-scale environmental projects can be turned into BSTEs, where learning is enhanced and monitored (this would be a form of action research), by way of introducing a context, a vision or an environmentally driven new technological component.

A BSTE can provide an opportunity for testing the feasibility of a new technology or service before it is ready to enter the open market. It can develop and test new social arrangements among actors and consider them as templates for other societal contexts. It is also a vehicle for drawing into the sustainability agenda actors who would otherwise not see a place for themselves in the types of projects in technological and system innovation that are often sponsored by powerful corporate, governmental, or NGO entities. A successful BSTE creates a functioning, socially embedded new configuration of technology or service that serves as a starting point for diffusion. An obvious indication that a BSTE is successful is when this new configuration first meets the initial expectations and then is widely replicated and becomes a social, environmental and commercial success.

Another, less obvious (and harder to demonstrate empirically), measure of the success of a BSTE is the occurrence of higher-order learning among its participants, even in the absence of wide replication. By this we mean one or more of the following occurrences:

- Participants re-examine, and possibly change, their initial perspectives on the societal needs and wants the project seeks to meet as well as the approaches and solutions

- Participants examine and place the particular project in a broader context of pursuing a sustainable society

- Participants examine, and possibly change, their own perceived roles in the above problem definitions and solution

- Participants change views on the mutual relationships among each other relative to the specific project or the broader societal context, including mutual convergence of goals and problem definitions

- Participants change their preferences about the social order as well as beliefs about best strategies for achieving them

A third indication of an experiment's success is a change in interpretive frames or problem definitions among two social groups: the future users of the new technology or service—the consumers—and the communities of practice represented by the participants in the experiment. Rohracher and Ornetzeder (2002; Ornetzeder and Rohracher 2006; Rohracher 2003) studied the impacts of participation in the design of, and living in, 'green' buildings on their occupants. Their studies have consistently shown that those future owners of residential units who participated in the planning, design and construction decisions would in the future interact more positively, and more effectively, with the energy-saving technologies. The experience also impacted the occupants' views on the issues of energy and environment and made them more receptive to trying out new energy-saving technological innovations. No such impact occurred among tenants who did not participate in the planning, design and construction decisions. Social theorists such as Storper (1996), Luthans and Kreitner (1985), Granovetter (1973), Bandura (1977) and Hamblin *et al.* (1979) as well as theorists of technological diffusion such as Rogers (1985) emphasise both the cognitive and the social processes involved. The cognitive component includes reflection, reassessment and reframing, as summarised in the preceding section, while the social component entails transmission and diffusion of new ideas and knowledge: in this case from the experiment's participants to the communities of practice or from the users of the innovation to their social milieus.

BSTEs, as defined earlier, have several characteristics that are conducive to higher-order learning among their participants. The presence of heterogeneous actors who represent different organisations, communities of practice and institutional affiliations assures the presence of a range of interpretive frames and belief systems. Moreover, the very act of choosing to participate in the experiment suggests a willingness on the participants' part to interact with each other and with each other's interpretive frames. The vision of sustainability, which is the driving force for at least some participants, has the potential to provide a platform, or an umbrella, for reframing the clashing interpretive frames, should conflicts arise. By evolving around a specific tangible 'thing'—the innovative product or service—the project provides focus and a shared language for the discourse.

The present study builds on the work of Grin and Van de Graaf (1996a,1996b) to conceptualise learning processes in BSTEs. These authors applied Fischer's (1995) and Schön and Rein's (Schön 1983; Schön and Rein 1994) frameworks of multilevel discourse to examine the learning processes occurring during constructive (or interactive) technology assessment. Their underlying assumption was that different professional communities (or communities of practice) can collaborate on a joint problem—despite partaking in different interpretive frames and problem definitions—as long as they share each other's problem definitions (shared, common or dominant problem definition; Vergragt 1988) or at least accept each other's problem definition as legitimate (congruence).

Participants in BSTE bring with them diverse perspectives and competences, which in turn affect the meaning they attach to the project at hand and the ways in which they seek to contribute to the project and its outcome. Factors such as professional training, self-interest, socialisation through membership in political and professional groups as well as deeply held values and beliefs contribute to the variability. We group these differences into four levels (closely following Grin and Van de Graaf 1996a, 1996b):

- Problem-solving according to predetermined objectives

- Problem definition with regard to the particular technological–societal problem coupling

- Dominant interpretive frames (including appreciative systems, value systems and background theories)

- World-view

Discourse at the first level to solve a defined problem generally includes tools that the participant deems fit for addressing the particular problem, such as engineering analysis, cost–benefit analysis and risk analysis. The discourse proceeds primarily within the participating groups, but may be between them, and relies mostly on professionally accepted norms. Learning at this level does not involve reflection on the objectives of the project or questioning of the match between the social problem and the solution that the particular technology represents. This is first-order learning.

Discourse on the second level is a struggle or negotiation about problem definition and problem–solution couplings. For instance, professionals with a technical background are inclined to frame the problem as being technical whereas social scientists would develop a more social problem definition. Learning on this level is adjusting problem definitions to reach consensus or, at least, congruence.

Each of the groups has a fairly stable interpretive frame (level 3), often tightly linked to their world-views (level 4). Changes in the interpretive frame are less frequent than on the level of problem definitions but may happen in periods of crisis. Participants usually interact on the second and third levels. This is where the differences in problem definition, motivations for engaging in the project, individual interests and organisational missions, and perspectives on the particular technology become most clearly exposed and are most likely to confront each other. The nature and the extent of the resulting higher-order learning depend on the form that that confrontation takes and the way it is managed by the BSTE participants. Generally, changes in problem definition are more likely than changes in interpretive frames.

It is also at these two levels of interaction that participants confront their own commitment to the process and its goals. Some may discover that they are not willing to engage with other participants in a way necessary to propel the experiment forward or are not open to self-reflection. They are likely to quit. New members might also join, either to fill the emerging vacancies or attracted to the project, the interactive process and possibility of learning. A BSTE is therefore a dynamic configuration not only from the perspective of changing ideas but also from the perspective of changing membership.

Discourse at the fourth level rarely occurs, is unlikely to produce changes and is most dangerous for a collaborative project. This is because the views of this order are very stable within each participant group. Rather than closing gaps in deeply held beliefs an open discourse in this domain may lead to a deadlock. Of course, differing world-views do play a role in the overall process. They do so indirectly, by impacting the way individual participants interpret the meaning of the project *vis-à-vis* the private and public interests or by how they define a problem.

The next section describes the innovative approach to building design as a sociotechnical experiment that was intended as a vehicle of higher-order learning.

3.3 Empirical case study of high-performance building in South Boston

Data for this case study were collected through participatory observation, from July 2004 to March 2006, in project meetings, from interviews with individual team members and from documentary analysis.

3.3.1 Developer's vision and the challenge

The South Boston neighbourhood, although geographically very close to the centre of Boston, is strikingly separate from it. Framed on three sides by the Boston Harbour, and separated from the city centre by a highway and a stretch of industrial buildings and warehouses, 'Southie' has evolved into a self-contained blue-collar working community with its own traditions and culture (MacKenzie 2003). Its residents are for the most part descendants of Irish immigrants who flooded this area at the turn of the twentieth century, and many still feel strong ties to their cultural ancestry. Ethnic and neighbourhood pride have fuelled the sense of separateness (over the past century, many leading politicians in Massachusetts came from South Boston).

But the neighbourhood is changing. The rising real-estate market, the growing interest among young professionals in city living and the advantageous location—near downtown, the airport and the commercial and cultural amenities of Boston—are squeezing out the area's long-time residents. The gentrification process started in the 1980s—as conversions of industrial buildings into lofts—and has accelerated during recent years in the form of luxury condominiums for workers in the nearby financial district of Boston as well as suburbanites returning to urban living (so-called Bobos: Bourgeois Bohemians; Brooks 2001).

The project we describe is part of this trend. During the 1980s the developer in question converted a historic mid-1800s rum distillery building he owned into lofts, which he rents to artists. In 2003 he decided to build, next to the Old Distillery, an approximately 80-unit and 150,000 ft² elegant residential building that would include apartments, open lofts, art studios and galleries. The developer is an atypical member of the real-estate development community. With a doctorate in philosophy and with a personal history of activism in the Students for a Democratic Society movement during the 1960s and 1970s, he divides his time between real-estate development and social research and writing. He is committed to using his financial resources and creativity for the betterment of the society. Accordingly, the new building will minimise its use of fossil fuels and will price a number of its original lofts below market value, in effect having the wealthy residents subsidise the modest-income residents (mostly artists).

The developer's ambition was to innovate in three areas: product (the building), process (designing and constructing the building) and end-use (the life in the building). With regard to the product, he sought to deploy as many cutting-edge energy-reducing technologies as possible, including the architectural know-how on high-performance design, and to rely on renewable energy (biofuel) for co-generation of heat and power, so that the net consumption of fossil-based energy will be zero. The list of energy-efficient technologies initially included:

- Solar heating panels integrated into the roof

- Heliostats (redesigned) to bring light into the interior as well as atriums

- Mobile louvers and reflecting shutters to modulate daylight and to provide insulation

- A compact building design to reduce loss of energy from the building envelope

- Photovoltaic cells as well as solar thermal panels

- Co-generation of heat and electricity

- Use of waste vegetable oil from local restaurants as heating fuel (biofuel)

- A greenhouse, as well as tomatoes in the atriums

- Storage of well water and winter ice in cellar through overcapacity electricity at night

- Insulation with heat exchange

- Air conduit systems for heating and cooling

- Car-sharing services, electric and plug-in hybrids, fuelled by electricity from co-generation

- Solar energy trapping by a glass south wall and porch

Taken individually, these technologies may not all be that new; it is their convergence into a single large-scale residential project, and the process of incorporating them, that would be innovative.

With regard to the process, the developer sought to arrive at the final design of the building by assembling a heterogeneous team—architects, urban planners, engineers, solar experts, energy consultants, grass-roots promoters of biofuels and artist-tenants of the current building—and by setting in motion an interactive, vibrant, creative discourse. Underlying this approach was his belief in (1) the creative potential of interdisciplinary teamwork and (2) the social benefit accruing from collaboration between business, professionals, artists and grass-roots activists. The developer also planned to employ local residents—those who lose the most as a result of gentrification—as contractors and subcontractors. In his vision, the benefits of doing so would extend beyond providing employment. He sought to engage these skilled workers in the exciting process of innovation in building design and construction and thus to provide them with a new perspective on the role of their respective occupations in innovating for the environment.

With regard to the end-use—the life in the building—the developer envisioned an organic vegetable garden and a transportation link with the centre of Boston that would encourage use of alternatives to single-occupancy vehicles. To that end, the developer planned to work with Boston's transportation authority to provide an efficient bus system for the residents of the building and its neighbourhood and to consider other options, such as car-sharing services and electric bicycles and scooters. In this vision, the building would attract occupants of a certain class—'savvy, well-educated, well-off, "elite cognoscenti", and critical of the status quo'—who would share with the developer and each other a belief that affluence need not be equivalent to high energy consumption. The building and its occupants would ultimately become a model for innovation in technology, design, process and lifestyle. All the participants, from the building designers to its future occupants, would, it was hoped, acquire a fresh perspective on their own role and, through diffusion, pass that perspective on to others within their professional and personal circles of influence.

The innovative process of designing the building presented a significant challenge because it in effect sought to change the traditional power relationships between a developer, an architect and a team. In conventional building projects, the architect is in charge. Working within the general parameters set by the developer, the architect makes the key choices with regard to the design, materials, technologies, consultants and the constitution of the overall project team. This management model is acutely sensitive to efficient use of experts' time and is highly risk averse. The risk aversion is rooted in the large financial risks involved in residential construction, the very real threat of litigation in case the new technologies do not work as planned and the prominent role of the insurance industry in real-estate development. The result is that project teams stay together from one project to another, and disincentives to trying new designs and technologies are strong.

In contrast, in this project the developer is in charge, and he assembled a team of experts who had not worked with each other before. Moreover, he asked these team members (including himself) to put aside their egos and some professional norms and to work in an open-ended, interactive mode, which he later called 'friendly competitive', where a multitude of ideas would be put forward, jointly discussed and possibly adopted. These ideas were scrutinised by outside experts as well as inside the group, with the developer as final arbiter. In essence, they were being asked to view the contribution of their own expertise through the lens of the overall project rather than seeing the project through the lens of their own expertise. The personality of the developer was the greatest asset in favour of this unconventional process management scheme: he is a softly spoken individual with strong interpersonal skills, a superb networker and listener, is genuinely interested in technical analysis and, despite keeping strong reins on the project, has a decisively non-authoritarian demeanour. His vision of the building was another asset, as long as it was shared by the team members. Nonetheless, as we discuss below, it was not easy to create a coherent project team that was able to comply with the developer's ideas for the process and the product.

3.3.2 The evolution of the process and emergence of the Core Team

By the end of the design stage the project Core Team consisted of six individuals who were intensely engaged in the collaborative process of designing the new building:

Architect 3, Architect 4, Urban Planner, Energy Consultant 3, Staff Engineer and the Developer. We capitalise their names to denote their roles as actors in a drama of sorts, in which each was both a representative of a profession and an individual (with a name, personality and value system). The Core Team emerged after a one-and-a-half-year period of the design stage, and after a significant turnover.

In addition, the Developer established a residents' advisory group consisting of the artists living in the existing Old Distillery building, to provide input on the issues of aesthetics and lifestyle. This group did not participate in the deliberations of the Core Team but rather entered the process at the point when the most important decisions about the combination of design and environmental performance had been already made.

Early on in the project, the Developer engaged the Urban Planner for his knowledge of the South Boston neighbourhood, extensive links to the architectural community in Boston, expertise in building models, knowledge of environmental regulations and for his skills in navigating the complicated building permitting process in Boston. Planner was an accomplished engineer and architect in his own right. The Developer also hired Architect 1. The initial drawings by Architect 1 were sleek but disappointing, show-ing little sensitivity to the energy and environmental aspects of the project. Architect 1's idea of a 'green building' was first to design the building and then bring technical specialists to add the 'green' features, in particular photovoltaic cells. As this was funda-mentally at odds with the Developer's goal of integrative design, zero fossil-energy con-sumption and affordability (which photovoltaic cells do provide), this idea was rejected and Architect 1 left the project.

About that time, the Developer hired Energy Consultant 1 (a solar design special-ist), and further a specialist on heliostats and louvers, and a local grass-roots activist specialising in promoting biofuels. The activist worked on an on-site biofuel-fired co-generation facility with the in-house Staff Engineer, who was responsible for technical maintenance of the existing Old Distillery. Originally, the Developer saw the project as a collaborative effort between established business and grass-roots activism. However, the activist did not value business culture and, after the permit for the co-generation plant was secured, left the project, whereas the Staff Engineer continued in an advisory role to the Developer.

Architect 2, with a high reputation in constructing high-performance buildings in Boston, was attracted to the project because of the innovative nature of the proposed building. His ideas were indeed more promising than those of Architect 1. The major advancement was to settle on a bulky, energy-conserving four-storey or five-storey structure with a large L-shaped footprint. The concept of internal light shafts emerged at this point, to solve the problem of delivering light to the interior walls of the apart-ments. Heliostats (revolving mirrors on the roof) would reflect sunlight from the roof down through the light shafts and into the apartments. Four apartments on each floor would be served by each light shaft and by a stairwell. For an illustration, see Figures 3.1a and 3.1b.

Both Architect 2 and Energy Consultant 1 were asked to engage with Developer and with other technical experts in interactive open-ended, free-flowing brainstorming ses-sions. However, it soon became clear that Architect 2 was not willing to adapt to this

FIGURE 3.1   (a) A rendering of the existing 'Old Distillery' building (dark) and the projected new residential building (light); (b) The building complex as seen from the NW; on top of the roofs the heliostats are visible

(a)

(b)

unusual mode of producing a building design, to the loss of control over the project or to the demands on his time that this process made. Energy Consultant 1 also resisted the process, especially the loss of control and the fact that the Developer asked him to consider their firm's specialty, photovoltaic panels, as only one of several technological options.

The project was in disarray. The original idea, a 'vibrant multidisciplinary team' to solve the problems jointly, was not working out. The participants became more entrenched, 'nobody was interested in learning anything from anybody'. Both Architect 2 and Energy Consultant 1 left the project. However, the Urban Planner, after initial scepticism, bought into the interactive team process and became its strong supporter. And soon new actors entered the project.

Architect 3, who replaced Architect 2, is a well-respected elderly retired gentleman whose experience goes back to Bauhaus and Gropius. His philosophy is that architecture is the means of improving human condition and preserving human dignity, and thus a building must be designed so that it satisfies some particular human needs. He sees a building as an aesthetic solution to a particular problem. Architect 3 embraced the interactive interdisciplinary team effort. Although it was mutually understood that Architect 3 would not take the responsibility for the entire project, he made some key contributions to it: he changed the spatial orientation of the building and replaced the light shafts with much larger atriums.

The atriums are designed to bring daylight in all levels deep into the building. They also fundamentally redefined the aesthetics of the building by opening the possibility of growing bamboo plants from the ground level to the roof, which would provide a view of sorts to the interior windows, by introducing an idea of window boxes with flowers (or tomato plants) and by providing more privacy than the narrow light shafts. The Developer seized the opportunity by introducing the idea of semi-tropical gardens in these atriums, with flowers and birds year round. However, they raised a number of issues concerning the layout of the apartments and the issue of how much 'leakage' of heat and cooled air would happen through these atriums. It also added extra costs, which became an issue later in the design stage.

One could dramatically impact the relationship among the inhabitants of the building and between them and the community outside, depending on where the living room windows would be placed, the orientation of the garage and the main entrance and whether the apartments sharing the same elevators were also sharing each other's (fairly intimate) view across the atriums. At one extreme, the interaction among the occupants would be enhanced, leading to the emergence of a strong sense of a community among them, but in that scenario the building would turn a blind face to the street. At the other extreme, the opposite would be true. The views of team members differed, with the Urban Planner favouring the community engagement, and Architect 3 focusing on creating a community among the building occupants. While the final answer was a compromise on both counts, the debate led to a lot of reflection. Furthermore, it gave rise to a new idea about the future life in the building: that the ground-level space of the building would contain, in addition to art galleries and retail outlets, commercial rental spaces to attract progressive innovative businesses specialising in building design and technology. In this vision, the building would become a hub for the Boston area innovators. In summary, the design issues impacting on lifestyle were:

- The impact of the atriums: to counteract privacy loss through apartment layout against use as opportunity for community building

- The orientation of apartments: inward-looking to build community of residents against outward-looking to reach out to the neighbourhood

Architect 4 joined the project a year into the process, and became its chief architect, with Architect 3 remaining on the team. Architect 4 has a long-standing interest in high-performance architecture, knows well some of its more famous examples, such as the Beddington Zero-energy development in the UK,[2] and has strong ties with the engineering community in Boston. He is cautious about the current 'green' movement within the profession, concerned that it could reduce the consideration of a building's performance to numerical scoring (as in the LEED [Leadership in Environmental Design][3]) certification), and thus rob the design of both process and content, and, ultimately, of creativity. Architect 4, after a long and successful professional practice within established institutions in the UK, Africa and the USA, opened his own solo practice in Boston a few years ago. He was seeking more challenges, fewer constraints and more opportunities to apply architecture toward solving social problems. The South Boston project was a perfect match.

Architect 4 found an instant understanding with the Developer about his values, personal mission and the view of this project. The interactive team process looked to him both intriguing and exciting. Having two architects on the team seemed more like a source of creativity than a threat.

The last two persons to join the team (during the second year) were Energy Consultant 3 and the Project Manager. The Developer brought in Energy Consultant 3 on the urging of the then Core Team members, who all agreed that the team needed someone to integrate the various technological and design ideas into a coherent whole. Energy Consultant 3 followed on the footsteps of departed Energy Consultant 1, and, for a brief and unremarkable time, of Energy Consultant 2.

Energy Consultant 3 is also a member of a non-profit organisation in Boston that is dedicated to transforming the residential building industry and thus furthering the goals of sustainability. The organisation concerns itself with the environment and with social and economic justice; it accounts in its designs for energy efficiency, environmental footprint, affordability and quality of residential construction. Energy Consultant 3 shared with the Developer and Architect 4 the key aspects of their vision, such as zero use of fossil fuels, providing affordable housing and challenging the dominant practices in the design and construction of residential housing. In the past, he has been a strong supporter of an interactive process such as the one applied in this project.

The Core Team emergent in the second year thus consisted of the Developer, Architect 3, Architect 4, Energy Consultant 3, the Urban Planner and the Staff Engineer,

2 en.wikipedia.org/wiki/bedzed
3 www.usgbc.org

with several additional experts and advisors for specific issues, as needed. The Core Team continued to meet once or twice a month for long discussions, with follow-up and more focused work in smaller configurations. Gradually, the meeting style changed from democratic and egalitarian to managed interaction under a clear leadership by the Developer and Architect 4. Notably, in that process, the Developer became highly conversant in all the technologies and their functions as well as in their role in the overall life of the future building.

The financial viability of the project became a central issue during the second year. Uncertainty in the real-estate market and the difficulty in estimating the final costs of the entire project led to various strategies. One was to build the new development in two phases: phase 1 would incorporate some new technologies and provide possibilities of learning along the way, and phase 2 would realise cutting-edge technologies and the atriums. This led to discussions about the permitting strategy: should permits be acquired for the two phases separately or jointly? Another idea was to rent, rather than sell, the units for the first few years, thus minimising the chances of litigation by owners in case glitches in the technologies should occur.

At that time, the Core Team also produced a value statement for the project. The Statement of Values is an interesting document in that it represents the Core Team's collective vision for the project. It was intended as a team-building instrument and as a platform for communicating with all the stakeholders, both internal and external to the project (including the permit-issuing agencies, politicians, marketing agents and so on). It would also be a guide for critical decision-making and conflict resolution among all the firms and individuals who would construct the building. The Statement also embodied the collective learning process that took place over the year and a half period of time during which the Core Team emerged and found its identity as a creative and effective group that set out to create an innovative building with a social mission.

Toward the end of the second year the main decisions about the environmental performance of the building, and the associated design, were made and the project entered the permitting stage. It is at this point that the Developer asked the residents' advisory group of the artists living in the existing Old Distillery building to become involved in the remaining design decisions. By that point the Core Team had *de facto* dissolved, so the residents' advisory group interacted primarily with the Developer and, through him, with Architect 4 who would integrate the new design features into the overall design. It focused on the aesthetic and lifestyle aspects of the building. One of the areas of consideration was the integration of public and private spaces. The Developer and the residents' advisory group would consider various scenarios of the movement of residents during different times of the day, days of the week and seasons (to consider Boston's hot and humid summers, cold and blustery winters, and mild autumns and springs), with an eye on designing the public spaces within the building and its external surroundings.

The overall goal was simultaneously to maximise the sense of community within the building, protect privacy and reaching out to the neighbourhood. Specific issues in this discussion included creating the space for a café, art galleries, commercial spaces, bus waiting areas and gardens. Another area of consideration was the affordability of the units, especially the wing of the building that would be occupied by the artists, but increasingly also (in the slower real-estate Boston market) all other units. Here, the residents' advisory group had an input into combining elegance and functionality with inexpensive materials, generally atypical for the construction industry. Other considerations included the design of the façade of the building that would integrate the existing nineteenth-century structure with the new construction, and integrating the primarily residential part with the work–life part that would be occupied by the artists.

Notably, our interview with the representative of the residents' advisory group indicated that this group was only superficially knowledgeable about the environmentally driven technical aspects of the building: 'we want [the Developer] to succeed because we want to show the world that it is possible to create an affordable, beautiful and environmentally highly performing building through private initiative' remarked an artist, but when asked about the details of the technology, she referred us to the members of the design team. It was clear in the interview that the artists' evidently strong sense of pride and shared ownership in the project was fuelled by their admiration for the Developer and his social mission, not by their own deep interest in ecology.

3.3.3 Analysis of the learning processes

The innovative nature of the South Boston building lies, paradoxically, in its conventionality: the greatest energy savings accrue from the bulky compact shape and large size. Energy Consultant 3 told us about his surprise in finding that the aspect ratio (the ratio between surface and content) generated by a computer model was that low. He did not think about it when first approaching the project. Together with external insulation this low ratio saves a maximum of energy.

The atriums are an old architectural concept, though certainly their function in this building has a new meaning, especially in combination with the heliostats. Neither is co-generation new, although the combination of co-generation with biofuels and with residential construction is. Similarly, the greenhouse and the elevated courtyard contribute to community building and the aesthetics but are not radically new features. It is the combination of all the above features in a large residential construction, on a risky and tight budget, that makes this project unusual.

The implementation process and the goal of the project are innovative, if not unique. A team-centred interactive process, driven by a value statement, has been applied elsewhere (as Energy Consultant 3 noted in the interview); but the Old Distillery project team was larger, and the duration of the deliberations (a year and a half) was longer by far than the usual practice. The goal of the project was to integrate cutting-edge environmental technologies and design know-how in a synthetic and innovative way in a residential building that would both be a commercial success and perform a social mission. The social mission consisted of:

- Reducing consumption of fossil fuels

- Challenging the conventional practices and norms within the architectural and construction professions and in the emerging specialty of so-called 'green architecture'

- Creating a replicable model of a zero-energy building, and of a collaborative team-based process of designing it, which would include different professional and occupational communities of practice

- Creating a replicable model of lifestyles in which wealth is not synonymous with a large environmental footprint

In short, this project fits the definition of a BSTE we gave earlier, and the process it followed—interactive, heterogeneous, focused on a specific goal and with a sense of urgency created by high financial stakes—was highly conducive for higher-order learning.

Building an effective team capable of pursuing the technical, economic and social objectives of this project took some time and a substantial turnover among the participants. In essence, the final chief architect (Architect 4) and Energy Consultant 3 stayed with the project because they fundamentally shared with the Developer and each other its objectives and the larger mission. Along the way they underwent a learning process, and so did other members of the Core Team. Below, we analyse the learning processes that took place by the four most active members of the Core Team in the course of this experiment, using the four-level conceptual scheme that was introduced in Section 3.2.

Table 3.1 summarises the views and beliefs of Architect 4, the Urban Planner, Energy Consultant 3 and Developer at the start of their participation in the experiment. The first thing to note about Table 3.1 is that three of the Core Team members have very similar world-views (level 4). They believe in the power of technology, creativity and collaboration between business and other societal actors in producing social good, they see the need for pursuing social justice through urban development projects and are committed to the goal of sustainability.

Table 3.1 shows that differences in interpretive frameworks among the four individuals, though not great, led nonetheless to notable differences in problem definitions. In the beginning of the project, the Developer focused on maximising the number and variety of ideas and on somehow turning them into a coherent building design through a democratic, egalitarian, interactive and inclusive process. This process, rooted in the Developer's views on authority and democratic participation, turned out to be too open-ended to produce progress, even for those team members who stayed with the team and was later replaced with a disciplined core group effort under the leadership of Architect 4. In an interview, the Developer mused that 'nothing I have learned ever presented authority and discipline as anything more than an atavistic issue of a bygone era . . . [I learned] that unless there are secure boundaries, firm rules and the guarantee of sanctions, dialogues and free participation are impossible.'

The Developer also discovered that designing a building with a radically better energy performance is an integrative process all the way through. At the outset, he

TABLE 3.1 Four-level learning scheme for the key members of the Core Team

Developer	Urban planner	Architect 4	Energy Analyst 3
Level 1: problem-solving			
Assemble, motivate and manage a heterogeneous team. Synthesise technologies and design while balancing competing objectives	Move the design process along towards successful approval and permitting	Move the design process along while allowing maximum creativity	Understand the conceptions of other team members. Analyse and optimise the energy flows of many alternative designs
Level 2: problem definition			
How to generate a multitude of ideas through interactive brainstorming within a shared vision and through an inclusive egalitarian group process. How to integrate as many innovative technologies as possible into a coherent design	First: how to get to know the community and institutions then meet their requirements within the Developer's parameters. Later, how to design a sustainable project, then how to articulate it to the community and institutions	How to integrate ideas emerging through an interactive process into a coherent and cost-effective design, guided by a shared vision	How to optimise the energy management of a building while integrating a multitude of other ideas and meeting competing objectives
Level 3: dominant interpretive frame			
Existing technologies can produce a radically different building: energy-efficient, affordable, beautiful, enhancing neighbourhood and lifestyles. Current professional practices impede innovation. Project-related egalitarian and inclusive collaboration of professions and communities of practice leads to innovation	Urban development should be integrated with community needs. The community and local institutions can impede real-estate development projects. Business can be advanced by satisfying their needs	Existing technologies can produce a radically different building: energy-efficient, affordable, beautiful, enhancing neighbourhood and lifestyles. The distinction between design and 'green' design is false. Current professional practices impede innovation. Architects are good synthesisers of ideas about technology, design, sustainability and human needs	Energy efficiency, high quality, and affordability are key features of sustainable design. Current professional practices impede innovation. System thinking and integration are key to sustainable design
Level 4: world-view			
Business, professions, technology and civil society can collaboratively produce change towards environmental sustainability and social equity. Democratic participation can take the function of strong authority	World-view unknown	Business, professions and technology can collaboratively produce change towards environmental sustainability and social equity	Business, professions and technology can collaboratively produce change towards environmental sustainability and social equity

rejected what he perceived as an unfortunate practice of designing a building and then adding to it 'ecological features', such as photovoltaic cells and other features. But, inadvertently, he landed in another unworkable extreme mind-set: assembling a list of innovative technologies and seeking ways to deploy them. The process of designing the South Boston building has taught him that technological features and the design must co-evolve.

Architect 4's problem definition accounted at the outset for a co-evolutionary approach to design. He did view the function of environmentally sensitive technologies as solving particular design and performance objectives, not just making the project 'green'. This problem definition flowed from Architect's 4 rejection of the current trend toward identifying 'green architecture' as a separate professional specialty and his broader view of his profession (as noted in the interpretive framework). What Architect 4 learned during the project was a deeper appreciation of the creative power of a collaborative, interactive, interdisciplinary team, given sufficient time. A conventional architectural design competition would not have created this level of open-ended interdisciplinary creative interaction.

Architect 4 also came to see new opportunities for the architect profession. In an interview, he described the new pressures on the profession in the USA, as builders and building suppliers increasingly encroach on what was once strictly the architect's territory: generating technical drawings, creating design ideas and thinking of the aesthetics. In his future vision, architects will let go of the old idea of being the only source of the aesthetic and creative part of the design and of being solely in charge of the entire project. Instead, they will engage in a collaborative interactive process with other relevant professionals and make powerful use of their powers of lateral thinking and synthesis. Architect 4 also came to think about this project through the Developer's lens: as a replicable model of a different building, different process of designing a building and a different lifestyle among its occupants. By doing so, Architect 4 moved closer to the Developer's social mission.

The Urban Planner's problem definition changed. A veteran in the Boston area real-estate development, he saw initially the design process as being strongly guided by the requirements of the neighbourhood and local institutions, while also satisfying the Developer's goals and the architect's ideas. By the end of the design process the Urban Planner began to see all these steps and (possibly competing) objectives in a more integrated way: a sustainable building design takes place through an interactive process that accounts for all perspectives and is therefore attractive to the neighbourhood and institutions. The process of obtaining a building permit consists of articulating the project's vision and its product: the design.

Energy Consultant 3 viewed this project as a technical challenge, a symbol and an instrument for advancing the cause of technological change toward sustainability. Through it, he sought to demonstrate that an 80% reduction in energy consumption can be done with current mainstream technologies and in the very difficult Massachusetts climate, which ranges from extreme winter cold and dryness to extreme summer heat and humidity ('if it is done, then it is possible', he mused). He sought to create a replicable model and to teach the construction professionals, through empirical experience and subsequent diffusion of knowledge, about high-performance buildings. He further aimed to raise the departure point for further incremental innovations in high-performance and environmentally sustainable construction.

His specific initial contribution to the project was to balance the Developer's enthusiasm for incorporating as many innovative technologies as possible with the considerations of cost, effectiveness and priority setting. Since the (bulky) shape of the building reduced the heating requirements substantially, the considerations of passive solar—rather standard in high-performance design—became much less. Instead, the analysis focused on efficient cooling systems (alternatives included groundwater circulation, aquifer storage, annual ice storage and diurnal ice storage), cost containment, equipment efficiency, fuel availability for the co-generation plant and maximising the residents' access to the attractive atriums. Other issues included the design of underground parking and the greenhouse and the designs of the roof heliostats.

The problem definition of Energy Consultant 3, focused as it was primarily on the energy aspects of the building, followed from his interpretive framework. It also accounted for the team process, which he entered when it was already well established. Notably, Energy Consultant 3 did not become the main force in integrating all the innovative technologies. This role fell on the Developer, with the help of many advisors. Rather, the focus of Energy Consultant 3 remained on the energy flows in the building.

Notably, Table 3.1 does not include the residents' advisory group because its members did not participate in the collective process of mutual interaction. Furthermore, the later addition of the residents' advisory group to the design process both limited its deep engagement in the process, including an enforced re-examination of problem definition, and precluded us from following the learning processes that would emerge. But this was not necessary. The interviews with two representatives from the residents' advisory group made it clear that they did not experience higher-order learning.

3.4 Social learning through bounded sociotechnical experiments toward sociotechnical transitions

The case study in South Boston shows that a BSTE can indeed induce higher-order learning among its participants. The key factors contributing to the learning include:

● The presence of a clear focus and boundaries for the project (to create a building)

● Intense and sustained interactions of several professionals with a commitment to the process and the goals of the project

● Agreement among the participants about the vision for the project and its social mission and about the process

● Agreement among them about the core social values, and overlap among the participants' interpretive frameworks

These factors (and the actual technical issues) constituted a bedrock on which the project participants could interact, solve problems, reflect on their individual interpretive frameworks and make changes in individual problem definitions. The availability of adequate time and funding also greatly helped. We also found that the participants in this interactive process who did not experience all of the above learning factors tended to leave the project.

The absence of higher-order learning among the residents' advisory group further underscores the fact that a sociotechnical experiment must have certain essential elements, as described above, to induce higher-order learning. Participation is not enough; interaction and feedback in a particular context must be present. Rohracher (2003) makes similar observations with regard to higher-order learning about the use of specific energy technologies in buildings: Rohracher recommends enhanced interaction between designers, developers and users (building residents) in the design stage of the building and its performance technologies.

This case study also identifies two units of analysis for studying team-based learning processes: the individual and the team. On the team level, the learning involved a gradual formation of a team that had the capability to carry out the sociotechnical experiment envisioned by the project champion. As it turned out, the experiment required that its participants achieve wide agreement on the fundamental values and interpretive frameworks. The team composition thus kept changing until this condition was satisfied. This observation is analogous to Schön and Rein's (1994) observation that intractable policy controversies arise from clashes between the contending parties' core world-views and interpretive frameworks. To resolve such controversies, these authors recommend reframing the problem in a way that eliminates the clash.

In our case, the conflicts were at first eliminated when the initial participants were replaced with others whose values and interpretive frameworks allowed congruence. The newly emerging Core Team resolved the remaining issues by multiple interactions and discussions, reframing and reformulating problems and collectively seeking solutions, each from an individual perspective and expertise. The Statement of Values was the embodiment of these fundamental agreements. The members of Core Team also agreed that the Value Statement, introduced early in the process, could be a powerful tool for selecting a design team and a construction team.

From the practical perspective, this case study shows that we must think of innovation in building design as both a process and a product. That means that when we want to replicate this building, we must ask two questions:

- What is it like and what features does it have?

- How was it developed and by whom?

This is a fundamental lesson about sociotechnical innovation through small-scale experiments.

Another practical lesson is to demonstrate the limitations of the model of learning we use here. Our case study, and our findings with regard to learning processes, applies to individual professionals working in a team toward a well-defined goal and under specific social and market conditions. Under these conditions it was possible for the team to change its membership until it achieved a shared vision, in this case a necessary condition for the learning to occur. Clearly, the model we used has limited applicability for groups of organisations working together, as such team turnover is hardly an option in those circumstances.

Finally, will the higher-order learning that took place in this BSTE diffuse beyond its boundaries? While the scope of our research cannot provide an empirically based unequivocal answer, several factors suggest that it will. First, the Core Team members have been energised by three ideas: that the building will serve as a model to emulate; that the design and construction of a sociotechnological innovation can and should be driven by a social mission and an explicit statement of values; and that the technology, the know-how and the professional capacity exist to propel the existing sociotechnical system of residential building construction towards a major shift. By their own admissions, they intend to carry these ideas 'back' to their respective communities of practice. Indeed, Architect 4 is, at the time of this writing, about to assume the presidency of the regional society of architects and has rich plans to influence his profession accordingly through that post. An interesting issue is whether the BSTE can become 'a story' that is a typical case used by others in disseminating levels 3 and 4, which is one of the initial objectives of BSTE.

Second, market forces in the design and construction industry facilitate the diffusion. Both Architect 4 and Energy Consultant 3 noted that individuals and firms who have in their portfolios 'green buildings' use them as a competitive advantage in typical bidding competitions. This project, if brought to completion, will raise the bar on what counts as an innovative 'green' building, thus giving the participants an additional competitive advantage.

Third, an important new idea emerged about the future life of the building: to attract innovators in environmental technologies and building design to the commercial rented spaces of the building. The first such business, the team member specialising in heliostats, is now renting a space in the existing Old Distillery building and will eventually move to the new building. Such a physical convergence of innovators will have a synergistic effect on the innovation process on the small and large scales alike. The potential for social change from creating such a critical mass, including the building occupants and resident artists, is hard to overemphasise.

Additionally, this building presents an opportunity for putting in place a mechanism for continuous learning and diffusion of ideas through feedback loops between the future building residents, building designers, providers of the technologies and other institutional players. The local authorities could play a key role in organising such feedback and disseminating its results. In the long run, this could be a very effective mechanism for social learning. It is encouraging that the local authorities in Boston are keenly interested in this building as well as the innovative transportation solutions planned for its future use.

3.5 Conclusions

This chapter presents a framework for monitoring higher-order learning in a hetero-geneous team of professionals and for enhancing such learning in particular settings, which we denote as bounded sociotechnical experiments. We studied a residential zero-energy building case in minute detail to obtain a deep understanding of the conditions under which learning does and does not take place. Our major findings are that the type of learning that has the potential to affect deeply entrenched professional prac-tices among building designers and developers can be facilitated; that it takes place both on the level of the individual and on the level of the team, by turnover of team members; and that inclusion of future building residents in the building design stage represents an underutilised opportunity for learning and for diffusion of ideas beyond the case in which it occurs.

Finally, this study exemplifies again that technological innovation is not only (and mainly not) about technology, but about people, their perceptions and their interac-tions with each other and with the material world and that sustainability will not be reached by technology alone but by deep learning by individuals, groups, professional societies and other institutions. Without higher-order learning sustainability will not be feasible.

References

Argyris, C. (1977) 'Double-loop Learning in Organisations', *Harvard Business Review* 55.5: 115-25.
—— and M. Schön (1978) *Organisational Learning: A Theory of Action Perspective* (Reading MA: Addi-son-Wesley).
Bandura, A. (1977) *Social Learning Theory* (Englewood Cliffs, NJ: Prentice Hall).
Berkhout, F., J. Hertin and A. Jordan (2002) Socioeconomical Futures in Climate Change Impact Assess-ment: Using Scenarios as "Learning Machines" ', *Global Environmental Change* 12: 83-95.
Birkland, T. (1997) *After Disaster: Agenda Setting, Public Policy, and Focusing Events* (Washington, DC: Georgetown University Press).
Brooks, D. (2001) *Bobo's in Paradise: The New Upper Class and How They Got There* (New York: Touch-stone).
Brown, H.S., and C. Carbone (2006) 'Social Learning through Technological Inventions in Low-impact Individual Mobility: The Cases of Sparrow and Gismo', *Greener Management International* 47: 77-88
——, P. Vergragt, K. Green and L. Berchicci (2003) 'Learning for Sustainability Transition through Bounded Socio-technical Experiments in Personal Mobility', *Technology Analysis and Strategic Man-agement* 13.3: 298-315.
Easterby-Smith, M. (1997) 'Disciplines of Organisational Learning: Contributions and Critiques', *Human Relations* 50.9: 1,085-113.
Fischer, F. (1995) *Evaluating Public Policy* (Boulder, CO: Westview Press).
Glasbergen, P. (1996) 'Learning to Manage the Environment', in W.M. Lafferty and J. Meadowcroft (eds.), *Democracy and the Environment: Problems and Prospects* (Cheltenham, UK: Edward Elgar): 175-212.
Granovetter, M.S. (1973) 'The Strength of Weak Ties', *American Journal of Sociology* 78.6: 1360-80.

Grin, J., and H. Van de Graaf (1996a) 'Technology Assessment as Learning', *Science, Technology and Human Values* 20.1: 72-99.

—— and H. Van de Graaf (1996b) 'Implementation as Communicative Action: An Interpretive Understanding of Interactions between Policy Actors and Target Groups', *Policy Sciences* 29: 291-319.

Hallsmith, G. (2003) *The Key to Sustainable Cities: Meeting Human Needs; Transforming Community Systems* (Gabriola Island, Canada: New Society Publishers).

Hamblin, R.L., J.L. Miller and D. Eugene Saxton (1979) 'Modelling Use Diffusion', *Social Forces* 57: 799-811.

Hoogma, R., R. Kemp, J. Schot and B. Truffer (2002) *Experimenting for Sustainable Transport: The Approach of Strategic Niche Management* (London: Spon Press).

——, M. Weber and B. Elzen (2005) 'Integrated Long-term Strategies to Induce Regime Shifts toward Sustainability: The Approach of Strategic Niche Management', in J. Hemmelskamp and M. Weber (eds.), *Toward Environmental Innovation Systems* (Berlin: Springer): 209-36.

Kemp, R., and J. Rotmans (2005) 'The Management of the Co-evolution of Technical, Environmental and Social Systems', in J. Hemmelskamp and M. Weber (eds.), *Innovation Systems towards Sustainability* (Berlin: Springer): 33-56.

Keohane, R.O., and J.S. Nye (1989) *Power and Interdependence* (Boston, MA: Scott, Forsman, 2nd edn).

Lee, K.N. (1993) *Compass and Gyroscope: Integrating Science and Politics for the Environment* (Washington DC: Island Press).

Luthans, F., and R. Kreitner (1985) *Organisational Behaviour Modification and Beyond: An Operant and Social Learning Approach* (Glenview, IL: Scott, Foresman & Company).

MacKenzie, E. (2003) *Street Soldier: My Life as an Enforcer for Whitey Bulger and the Boston Irish Mob* (Hanover, NH: Steerforth Press).

Ornetzeder, M., and H. Rohracher (2006) 'User-led Innovation and Participation Process: Lessons from Sustainable Energy Technologies', *Energy Policy* 34: 138-50.

Paquet, G. (1999) *Governance through Social Learning* (Ottawa: University of Ottawa Press).

Portney, K.E. (2003) *Taking Sustainable Cities Seriously: Economic Development, the Environment, and Quality of Life in American Cities* (Cambridge, MA: MIT Press).

Register, R. (2002) *Ecocities: Building Cities in Balance with Nature* (Albany, CA: Berkeley Hills Books).

Rogers, E.M. (1985) *Diffusion of Innovations* (New York: The Free Press, 4th edn).

Rohracher, H. (2003) 'The Role of Users in Social Shaping of Energy Technologies', *Innovation* 16: 177-93.

—— and M. Ornetzeder (2002) 'Green Buildings in Context: Improving Social Learning Processes between Users and Producers', *Built Environment* 28.1: 73-84.

Sabatier, P. (ed.) (1999) *Theories of the Policy Process* (Boulder, CO: Westview Press).

Schön, D.A. (1983) *The Reflective Practitioner: How Professionals Think in Action* (New York: Basic Books).

—— and M. Rein (1994) *Frame Reflection: Towards the Resolution of Intractable Policy Controversies* (New York: Basic Books).

Senge P.M. (1990) 'Building Learning Organisations', *Sloan Management Review* 32.1: 7-23.

Sitkin, S.B. (1992) 'Learning through Failure: The Strategy of Small Losses', *Research in Organisational Behaviour* 14: 231-66.

Storper, M. (1996) 'Institutions of the Knowledge-based Economy', in *Employment and Growth in the Knowledge-Based Economy* (Paris: OECD): 255-83.

Van Eijndhoven, J., W. Clark and J. Jager (2001) 'The Long-Term Development of Global Environmental Risk Management: Conclusions and Implications for the Future', in The Social Learning Group (ed.), *Learning to Manage Global Environmental Risks. Volume 2: A Functional Analysis of Social Responses to Climate Change, Ozone Depletion, and Acid Rain* (Boston, MA: MIT Press): 81-97.

Vergragt, P.J. (1988) 'The Social Shaping of Industrial Innovations', *Social Studies of Science* 18: 483-513.

Wenger, E. (1998) *Communities of Practice: Learning, Meaning, and Identity* (Cambridge, UK: Cambridge University Press).

—— (2000) 'Communities of Practice as Social Learning Systems', *Organisation* 7.2: 225-46.

Wildawski, A. (1990) 'Choosing Preferences by Constructing Institutions: A Cultural Theory of Preference Formation', *American Political Science Review* 81: 3-21.

4

S-House: sustainable building utilising renewable resources— Factor 10 building with innovative solutions

Robert Wimmer and Myung-Joo Kang

Centre for Appropriate Technology, Austria

In 1999, the Austrian Federal Ministry of Transport, Innovation and Technology (BMVIT)[1] launched the research and technology programme, Sustainable Development, which aimed at effective stimulation of a restructuring of the economy towards a sustainable future. Various research and development projects as well as demonstration and diffusion measures have been supported by a number of subprogrammes in order to give new impetus to innovation in Austria's economy.

The S-House (Sustainable Straw-bale House; see Fig. 4.1)[2] project was launched and carried out within one of the subprogrammes, the Building of Tomorrow, by the Centre for Appropriate Technology (GrAT)[3] at the Vienna University of Technology, and financed by the EU Life Environment Association and the State of Lower Austria.

The designing and building processes of the S-House followed the idea of resource circulation and took into account the entire life-cycle of materials and energy streams used throughout all phases of the product as well as all relevant consequences. From the research and prototyping experiments carried out earlier, it was clear that efficient and innovative utilisation of renewable raw materials is a key factor for sustainable building concepts.

---

1 Bundesministerium für Verkehr, Innovation und Technologie.
2 Wimmer *et al.* 2005.
3 Gruppe Angepasste Technologie.

During the project a wide range of innovations were initiated and realised, and new products based on renewable raw materials were developed. Among those materials, abundant natural resources such as straw bales were used in order to contribute to the enhancement of regional economic growth. Those building materials obtained from the region also ensure simple rebuilding and re-usability.

FIGURE 4.1 **North and south façade of the S-House**

4.1 Case description

4.1.1 Overview

In accordance with the criteria of sustainable building, the S-House project aimed to consider the following issues with regard to its concepts and components:

- Reduction of energy and material consumption
- Promotion of the use of renewable energy sources
- Use of renewable and ecologically sound raw materials
- Social aspects
- Improved quality of life
- Similar costs to those of conventional building construction

To reach these goals, a Factor 10 office building suitable for series production was planned and realised.[4] Innovative design and construction elements have resulted in a drastic reduction of energy and resource consumption compared with conventional

4 Factor 10 compared with conventional buildings.

building technologies. This enormous resource efficiency is intended to be fulfilled through the whole life-cycle of the construction and its use.

These goals have been achieved by an innovative combination of passive house technology, renewable resources and regionally available materials (such as straw-bale insulation, wooden construction and clay plaster) in a modular and contemporary architectural design. Technical performance of the demonstration is also constantly monitored.

4.1.2 Case context: landscape and regime

The building market is a business segment with a particularly large volume of material flows (Fig. 4.2) and high energy consumption resulting from manufacturing, transportation and deconstruction processes. Most conventional building components and structures are based on mineral and fossil raw materials such as metals, oil and uranium. These resources are in limited stock and cause severe environmental problems during their life-cycle. Also, large quantities of building waste create significant environmental impacts and have a correspondingly high cost for disposal.

FIGURE 4.2   **Material consumption by business sector for various Austrian industries, 2003 (metric tons)**

Source: Statistik Austria 2004

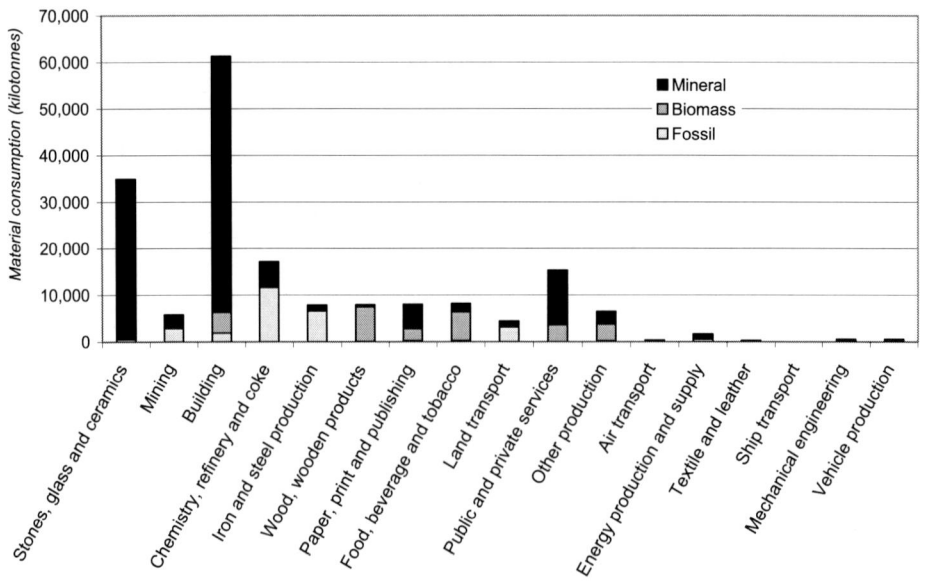

If we look at traditional local buildings throughout the world, they have common sustainability factors such as:

- Use of renewable materials (for example, bamboo, other plant material, clay)
- Use of regionally available materials
- A unique building structure that fits the climate and environment of the region
- Ease of build, operation, maintenance and deconstruction
- Provision of a healthy environment (for example, in terms of indoor air and light)

As a demonstration example of sustainable building, the S-House was carefully planned and constructed from the very beginning. The idea of resource circulation has been integrated along the entire life-cycle, from the selection of raw materials to end-of-life planning. In particular, the S-House project focused on the efficient utilisation of renewable raw materials, which have a decisive influence on the building system.

4.1.3 Actors and their roles and perspectives

Planning and realisation of the S-House was based on results obtained from research work conducted in cooperation with various partner organisations over the period of a year (Wimmer *et al.* 2001a). In the course of these fundamental research studies, the technical, legislative, political and organisational scope of utilising renewable raw materials in building systems were considered and tangible design solutions developed (Wimmer *et al.* 2001b).

4.1.4 Case history and development

4.1.4.1 Utilisation of renewable raw materials

To prove the feasibility of the S-House concept and to identify further development needs towards series production, extensive prototype experiments were undertaken. According to the results obtained from the fundamental studies, the façade was built as a structure made of wooden boards and straw bales that are pressed and mounted free of thermal bridging. This straw proofing was provided with a layer of clay plaster and a wooden casing (see Fig. 4.3). Thus, the entire building is 'packaged' with straw, providing optimum heat insulation. The physical properties, highly effective heat insulation and fire resistance, of the straw-bale construction were tested during extensive studies.

Investigations indicated that straw possesses excellent physical and constructional properties and that the tested wall construction easily achieves passive house standard because of its good heat insulating qualities (see Table 4.1).

TABLE 4.1 Technical details of the straw bale wall test for the S-House[5]

Item	Result
Fire resistance	F90
Flammability test[a]	B2, E
Thermal conductivity	0.0456 W/mK
Sound reduction	53 dB

[a] To OENORM B3800 (B2), EN ISO 11925/2 (E).[6]

Source: Wimmer *et al.* 2001a

FIGURE 4.3 Construction progresses: top left, wooden casing; bottom left, straw bales being manoeuvred into place; right-hand side, clay plaster cladding being applied

5 The fire resistance value of F90 means that escape routes beneath the fire are protected for up to 90 minutes. The minimum requirement for a detached house is F30. The minimum standard for sound reduction ranges between 38dB and 52dB for an office building.

6 Austrian standard OENORM B3800 was replaced by European standard EN 13501-1 on 3 May 2010.

4.1.4.2 Intelligent building structure and reduction of resource consumption

The building stands on a sub-ventilated building slab, which is supported by individual footings. This structure facilitated a significant reduction in the use of mineral resources as compared with a conventional foundation and does not allow accumulation of cold and moist air in the floor slab area. The building has a modular design that enables easy maintenance and a long lifetime.

Solar radiation energy is captured by the large glazed south façade and is distributed through a mechanical ventilation and exhaust system in the building. The air is transported by specially developed wooden channels into all areas of the building. In the ground floor, a stone floor acts as a heat retainer. The stone floor tiles show an outstanding heat storage capacity and are the only mineral material that is used in the S-House. The stone tiles are glued with a natural adhesive substance so that recycling is ensured even here. An earth commutate for the air (a pipe in the soil at a depth of 2.5 m where the temperature is always above freezing) takes care of temperature balancing: during winter it prevents ice formation in the ventilation system and during summer it serves as a cooling device. The centrally located (backbone) supply cable ensures very short conduction paths in the intermediate ceiling area for electrical power supply and illumination resulting in less cable length, less material consumption and easier maintenance of the wiring. A daylight control system ensures efficient operation of the illumination system.

Not only the overall building structure but also detailed elements were designed with careful consideration. One example is the Treeplast screw, specially developed for the S-House (Drack *et al.* 2004). Straw bales have an heterogeneous, rough structure. Ordinary nails and screws that can be used in conventional building are not suitable for the S-House in terms of their strength resistance or the sustainability philosophy behind the construction. The means of mounting other building items such as shelves and façade components to the straw-bale wall were investigated. To fulfil the aim of the straw-bale building, not only is the shape of a device important but also the material used for the device. The team therefore followed an ecodesign process to come up with an environmentally sound solution and succeeded in developing a unique straw-bale screw. The aim of the screw design was to achieve maximum mechanical strength with minimum material consumption. The screw has been optimised during its development according to the principle of biomimicry (Fig. 4.4).[7] As a result the screw was made of renewable resources (biopolymer), is biodegradable, allows easy dismantling of the construction and is re-usable.

---

7 Biomimicry is concerned mainly with the application of biological principles to technology. Many principles of design, methodology and development seen in nature can be applied in many fields, especially in the building sector.

FIGURE 4.4 Treeplast® screw made of biopolymer

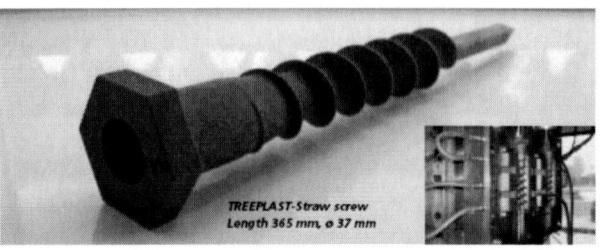

φ = the diameter of the screw shaft

4.1.4.3 Full utilisation of regional resources

By using raw materials (clay and straw) available on-site or in the region, environmental pollution caused by the manufacture and transportation of building materials is minimised.

Straw bales are by-products of cereal cultivation and their production life-cycle is very short. As a building material, straw bales have good availability and generate no extra cost in production.

Another innovative use of a regional resource was the collection of clay plaster for external application. The excavated material collected while preparing the individual footings and the installation shaft was stored so as to re-use it later. After the clay-containing earth material was separated from the humus layer and was made free from organic substances it was possible to process it into clay plaster for the walls of straw bales. The clay plaster applied directly on straw was an excellent alternative to commonly used films, which mostly compound material made of fossil-based and synthetic substances. This 'direct material recycling' indicates an important alternative to the conventional method of building where large quantities of ground excavation and building rubble and other leftover materials normally occur.

The surface of the roof is covered with a light layer of vegetation. This green roof protects the waterproofing membrane from UV radiation and helps rainwater absorption. The collected rainwater is used for watering plants, and cleaning.

4.1.4.4 User's comfort

Based on a patented system (Wimmer 2001, 2003, 2005) a biomass stove, Simplyfire, has been developed and integrated into the heating and air distribution system in order to cover thermal load peaks. This stove operates similar to the principle of ceramic tiled stoves, which means heat can be stored and delivered through the ventilation system at staggered timings. The unit, which is usually used as a stand-alone stove with radiation heat transfer, was adapted to the requirements of passive solar houses in two ways. First, the air used for burning is taken from the outside of the house, because the controlled ventilation must not be disturbed by stoves. Second, the generated heat is distributed to the whole house. Besides the technical innovations, another important objective of the stove was to take the users' needs into account with regard to their comfort. User surveys indicate that additional heating systems used to date in passive houses are often seen as a deficiency as there is no fire place and so there is an absence of a 'warm core' in the living room. This new biomass storage oven with a visor allows direct viewing of the heat source even in the passive house (Fig. 4.5).

In addition, the indoor climate is totally free from synthetic or chemical substances. The use of two types of special wood materials for ventilation system produces a pleasant atmosphere.

FIGURE 4.5 Simplyfire biomass stove

4.2 Results

4.2.1 Main result: achievement of Factor 10+

By using building materials derived from renewable raw materials and passive-house technology, the consumption of resources during the construction of a building can be minimised by Factor 10 compared with conventional construction methods. The comparison of a straw wall construction with a conventional wall construction has shown that the straw wall scores better by Factor 10 in all criteria, which means, for a conventional concrete wall construction the consumption of natural resources is 10 times greater. The consumption of resources during the use of the building could be minimised through innovative solutions up to a factor of 20. These calculations were made in cooperation with the Austrian Institute for Building Biology and Ecology (IBO).[8]

---

8 Österreichisches Institut für Baubiologie und Ökologie; also translated as the Austrian Institute for Ecological and Healthy Building (www.ibo.at/en [accessed October 2009]).

The construction of the S-House aimed to provide a modern, comfortable, high-performance office environment and at the same time spare the resources and minimise building leftover materials during construction of the building and to re-use all building components in order to avoid environmental pollution even after the usage period of the building. Owing to the special construction of the building (see Section 4.1.4.2, use of individual footings and sub-ventilated building slabs) even preparatory works at the building site could be done without excessive waste of building materials. The hollow spaces for individual footings were covered with film material after excavation in order to enable easy removal of the foundations. Each individual footing is prepared for easy removal and later disposal. Also, the vegetation and rainwater retention on the green roof has a very low environmental impact.

All performance levels have been recognised by the relevant authorities and the S-House has won many national and international awards such as the Architecture and Sustainability National Award (Staatspreis für Architektur und Nachhaltigkeit) in Austria,[9] the Rio Award in Germany[10] and the Global 100 Eco-Tech Award in Japan.[11] In 2009, it was nominated as one of the finalists for the international Index:Award in recognition of its positive impact for quality of life improvement.

4.2.2 Change in sustainability performance

4.2.2.1 Environmental performance: passive house criteria

The design-related solutions developed for the S-House meet all requirements of the passive house standard with regard to heat conductivity and wind proofing. All structural elements and designs offer a high degree of safety and user comfort because of the use of non-toxic building materials and ecologically sound design. On the whole, no metallic components or fossil-based synthetic materials were used in the building shell. Only wood was used for the static construction.

Owing to the optimum insulation of the building and the passive house technologies used, the S-House achieves a low energy consumption (6 kWh/m²a), which is even below the standard required for a passive house (15 kWh/m²a). The energy is supplied by solar radiation and biomass up to 100%.

4.2.2.2 Economic performance

Economic viability has been met at the highest level, thus showing the potential of the S-House as a feasible sustainable building design for series production. The calculated building cost is equal to that of conventional buildings. The costs of maintenance are very low. The annual heating cost is approximately €50 for firewood for the 400 m² office space. Multi-functional space concepts (such as movable interior walls) allow easy adaptation in case of future changes. Furthermore, de-installation costs will be

---

9   www.klimaaktiv.at/article/archive/27722
10   www.rio-innovation.de
11   www.expo2005.or.jp/en/eco/eco_tech.html

extremely low because of the use of biodegradable untreated materials and the re-usable house technology. The S-House demonstrates a successful example for sustainable innovations combining economic and ecological advantages.

4.2.2.3 Social performance

The use of straw bales is a new income source for the agriculture industry in the area. Farmers can sell this by-product from agriculture, which in the past had very little economic value, as a valuable building material.

One of the most important roles of these sustainable buildings is to give people the chance to gain access to tangible solutions and to ensure feasibility. Seminars, workshops and conferences offer the possibility to spread knowledge among professionals, politicians, students and self-builders.

4.2.3 Learning experiences

It is significant that all people involved in the project (scientists, architects, practitioners, material providers and so on) play an important role in the design and implementation processes. Suggestions based on their expertise were given serious consideration and were reflected in the harmony with other applications.

In order to monitor the environmental performance of the building, an extensive measuring system has been developed to estimate and document the important physical and climatic parameters of the building. Thus, the data obtained in the lab regarding the building construction are now being tested in practical terms.

On the north side of the façade, test wall elements are built into the straw and wood construction where additional insulating materials made from renewable raw materials are applied. This part of the façade is equipped with measuring sensors so that a comparison can be drawn between the wall elements and the rest of the building. For this purpose, insulating materials made of cork, hemp and flax were built in. Different measurements are continually being evaluated and analysed.[12]

Throughout the operation and maintenance of the building a great number of cutting-edge sustainable building techniques have been demonstrated. The S-House as an example of the combination of those techniques has drawn worldwide attention, and a great number of visitors have shown their interest and willingness to replicate the sustainable building concept.

4.3 Potential for diffusion and scaling up

Today the S-House is used as a centre for sustainable technology, renewable raw materials and sustainable building technologies. The components and design structures are

---

12 Product specifications and examples can be found at www.nawaro.com.

presented to visitors and therefore traditional know-how and recent development in the building industry are accessible to the public. In addition to the exhibitions, technical events such as symposiums, workshops and advanced learning sessions are organised (Fig. 4.6).

FIGURE 4.6 The S-House as an information dissemination centre

Among those events, a series of workshops on 'Modern Construction and Renovation with Straw Bales' for self-builders demonstrate straw-bale building techniques. Participants learn about the quality criteria of straw bales, examination procedures, the method of designing and planning foundations, wall and roof constructions and junctions between foundations.

This dissemination activity of the knowledge and experience gained from the project contributes to the growth of public awareness of existing alternatives to conventional building technology and gives participants a chance to learn the relevant skills (Fig. 4.7). In addition to the hands-on activities, information support is provided on the website.[13]

FIGURE 4.7 Participants in the practical workshop learning about clay and natural paint

13 See www.s-house.at.

The S-House also forms an important part of the theme park on 'Sustainable Technological Development' planned in the Böheimkirchen area, in which the building is constructed. Around the S-House there is a spacious 'material garden' which exhibits the natural building materials that are used in the construction as well as extended structures and suitable materials for ecological garden landscaping (fences, path, facings, etc.).

An international competition on sustainable architecture and design has been launched with the purpose of encouraging university students to come up with more creative utilisation of renewable materials and to develop the idea further with support of professional planners and architects (see www.grat.at/competition [accessed October 2009]).

One of the objectives of the project is to stimulate a series of sustainable buildings. In order to realise this aim, industrialised production methods will be further developed.

4.4 Overall conclusions

The power of the S-House project is to show tangible results based on research. Feedback from visitors often indicates that the experience of seeing this demonstration building that follows the important principles of sustainable development is very striking. Instead of asserting trite environmental messages the S-House shows the positive, pleasant and smart results that can be obtained from a different development option that really works yet has a low environmental impact. Many follow up activities from a broad range of stakeholders on an international level show that the concept has high potential for replication.

References

Drack, M., R. Wimmer, and H. Hohensinner (2004) 'Treeplast Screw: A Device for Mounting Various Items to Straw Bale Constructions', *Journal of Sustainable Product Design* 4: 33-41.

EN ISO 11925-2 (2002) 'European Standard: Reaction to Fire Tests: Ignitability of Building Products subjected to Direct Impingement of Flame. Part 2: Single-Flame Source Test' (Brussels: European Committee for Standardisation).

EN 13501-1 (2007) 'European Standard: Fire Classification of Construction Products and Building Elements. Part 1: Classification Using Data from Reaction to Fire Tests' (Brussels: European Committee for Standardisation).

OENORM B3800 (1997) 'Brandverhalten von Baustoffen und Bauteilen' ('Fire Resistance of Building Materials and Parts') (Vienna).

Statistik Austria (2004) 'Results of the 2004 Structural Business Statistics Manufacturing and Services', www.statistik.at (accessed October 2009).

Wimmer, R. (2001) Patent Number A407438, Austria.

—— (2003) Patent Number DE 19859353, Germany.

—— (2005) Patent Number CH 694471, Switzerland.

——, L. Janisch, H. Hohensinner and M. Drack (2001a) 'Wandsysteme aus Nachwachsenden Rohstoffen' (Wall Systems made of Renewable Resources), *Berichte aus Energie- und Umweltforschung* 31 (Vienna: Bundesministerium für Verkehr Innovation und Technologie).

——, L. Janisch, H. Hohensinner and M. Drack (2001b) 'Erfolgsfaktoren für den Einsatz nachwachsender Rohstoffe im Bauwesen' (Success Factors for Renewable Resources in the Building Sector), *Berichte aus Energie- und Umweltforschung* 24 (Vienna: Bundesministerium für Verkehr Innovation und Technologie).

——, H. Hohensinner, M. Drack and C. Kunze (2005) 'S-House Innovative Nutzung von nachwachsenden Rohstoffen am Beispiel eines Büro- und Ausstellungsgebäudes' (S-House Innovative Utilisation of Renewable Resources at an Office and Exhibition Building), Nachhaltig Wirtschaften konkret, *Berichte aus Energie- und Umweltforschung* 02 (Vienna: Bundesministerium für Verkehr Innovation und Technologie).

5

Rolf Disch's Solarsiedlung am Schlierberg: a solar housing estate in Freiburg; from architectural vision to entrepreneurial reality

Rolf Wüstenhagen

Institute for Economy and the Environment, Switzerland

Solarsiedlung am Schlierberg is an award-winning solar housing estate in the city of Freiburg im Breisgau, Germany. Rolf Disch, an architect with long-standing experience in low-energy architecture, has driven the project through a 10-year period that started with the planning of the first buildings in 1997 and resulted in construction and completion of 50 single-family homes and a commercial building complex between 2000 and 2006.[1] Today, 180 people are living in the solar housing estate, and the related commercial building provides office space for an additional 240 people.

Although low-energy architecture has recently become more and more popular in various European countries, this project is unique in that the buildings generate more energy than they consume, which is why Rolf Disch has coined the term 'plus-energy houses'.[2] This result has been made possible by combining the latest insights into energy efficiency in buildings (such as high-performance transparent insulation and heat recovery) with the replacement of traditional roof structures by large-scale photovoltaic panels. However, the scope of the project goes beyond a demonstration of smart architectural and technological innovation, as one of the ambitions (partly

---

1 See the websites Solar Siedlung GmbH: www.solarsiedlung.de (accessed October 2009)and www. rolfdisch.de (accessed October 2009).

2 See the Das Plusenergiehaus® website, at www.plusenergiehaus.de (accessed October 2009).

driven by necessity) was to provide affordable housing, leading to financial innovation such as the launch of a series of dedicated solar real-estate funds. The plus-energy houses are only 15% more expensive than conventional single-family houses; therefore the reduced energy consumption provides for rapid payback and, in all likelihood, increasing real-estate value over time.

An important element in surviving the ups and downs of the project was the project champion's move from being a pure architect to becoming a social entrepreneur, who also addressed the financial and regulatory challenges of realising a pioneering project of this size. Today, after successful completion of the project, the estate serves as a model that has inspired visitors from around the world to pursue similar activities in implementing solar architecture for residential and commercial buildings.

5.1 Case description

5.1.1 Case context: landscape and regime

Consumption of non-renewable energies in buildings is a major sustainability concern and results in a significant contribution to climate change. Therefore, improving energy efficiency in buildings and increasing the share of renewable energy are important elements of a sustainable future. This has led to various building-related energy policy targets. For example, the Swiss Federal Energy Research Commission (CORE) suggests a complete elimination of fossil fuels from the heating sector by 2050 (CORE 2007).

Experimentation with buildings that consume less energy dates back at least to the first oil price shocks in the 1970s. What used to be a matter of some visionary environmentalists designing and building avant-garde concept houses in the USA[3] and Sweden became a serious field of activity for architects and building physicists in the late 1980s and early 1990s. A particular milestone in Central Europe was Wolfgang Feist's so-called passive house, completed at Darmstadt, Germany. in 1991. By optimising several building elements, this multi-family house is characterised by an energy consumption of just 10 kWh/m²a, which was 95% less than the average consumption of typical residential buildings constructed at the same time. Rolf Disch entered the scene with his eye-catching Heliotrop rotating solar house completed in 1993, as well as with a couple of low-energy single-family houses completed at Freiburg in the same year. He brought a new focus to the low-energy house community, as his objective with these projects was less on record-low energy consumption but on finding the right trade-off between building cost and energy efficiency.

Although other pioneering landmark buildings followed at various locations in several European countries (such as the all-renewably powered manufacturing building

---

3 See, for example, the former headquarter of the Rocky Mountain Institute, built in 1984, which now serves as Amory Lovins's private residence (www.rmi.org/rmi/Amory's+Private+Residence [accessed 3 November 2010]).

of Solar-Fabrik AG in Freiburg completed in 1998), it is fair to say that low-energy and solar architecture remained a rather small niche, with mainstream construction markets following the trend towards higher energy efficiency only very slowly (plus larger and larger living areas provided for significant rebound effects). By the end of 1999, there were about 300 building units in Germany that met the specifications of a 'passive house'. By 2006, their number had increased to between 6,000 and 7,000 (Feist 2006), some 0.02% of German residential buildings. Similarly, about 6,500 houses in Switzerland have been built according to the Minergie-Standard for high energy efficiency, of which some 100 according to the strictest standard Minergie-P,[4] and 1,600 buildings with approximately 4,000 apartments have been constructed as passive houses in Austria.[5]

Some progress has been made on the regulatory level. Building energy efficiency standards in Germany have been tightened over several rounds since the initial introduction of a law on energy efficiency in buildings in 1976 (EnEG,[6] WSchV[7]). The two latest rounds of improvement are the Energy Savings Ordinance (EnEV) in 2002 and its recently introduced amendment in 2007 (EnEV 2007).[8] This latest update includes the introduction of energy performance certificates for buildings in 2008, which may increase consumer awareness about the issue and thus provide an additional incentive for accelerated diffusion of energy-efficient building practices. Finally, over the past two years, consumer interest in energy-efficient buildings has been sparked by the strong rise in oil and gas prices. So, although some important early indicators of change towards more sustainable building practices can be identified, a large-scale market penetration of such buildings has yet to materialise. Adaptation of regulatory standards appears to have long lead-times. Also, architects with low levels of awareness regarding energy issues, long reinvestment cycles and investors that stick to established networks of actors along the construction supply chain seem to inhibit the level of change that is required to meet ambitious policy targets.

5.1.2 Actors and their roles and perspectives

5.1.2.1 Primary actor: Rolf Disch

Against the mixed background described in the previous section, Rolf Disch started out to make a difference and create a model for affordable, high-efficiency residential buildings that would produce more energy than they consumed. An architect by training, his primary role at the outset of this project was to design and plan the buildings and then leave implementation to others. He was driven by a deep conviction about the importance of energy and environmental issues and an ambition to create a positive example. Over the course of the project, he realised that he had to transform himself

---

4 See www.minergie.ch (accessed October 2009).
5 www.igpassivhaus.at (accessed October 2009).
6 www.bmwi.de/BMWi/Navigation/Service/gesetze,did=22086.html (accessed September 2010).
7 de.wikipedia.org/wiki/W%C3%A4rmeschutzverordnung (accessed September 2010).
8 www.bmwi.de/BMWi/Navigation/Service/gesetze,did=23820.html (accessed September 2010).

from architect to social entrepreneur in order to make the project happen (see Section 5.1.3).

5.1.2.2 Other parties: investors and municipal authorities

The two other main important actors in the case were the investors and the municipal authorities of the city of Freiburg. The initial investor, Rolf Deyhle, looked like the ideal partner, with sufficient financial strength to help realise the project, but he had to withdraw in the late 1990s. Two philanthropic investors, Alfred Ritter and Marli Hoppe-Ritter, jumped in to fill the gap, but the change in investors caused the near collapse of the entire project. This was partly because of the risk-averse attitude of the municipal authorities, which decided to sell half of the land to a conventional real-estate investor after the reorganisation of the financial plan.

5.1.3 Case history and development

5.1.3.1 Early roots of the Passive House Movement (since 1973)

As discussed in Section 5.1.1, the idea of building houses with substantially lower energy consumption arose in the aftermath of the 1970s oil price shocks. For many years, it was constrained to a niche of visionary environmentalists who developed pioneering technological concepts. However, market implementation was limited to a small number of buildings.

5.1.3.2 Visionary ideas (1993–96)

Like many of his pioneering peers in the field of sustainable housing, Rolf Disch started out with some experimental projects, which included his private house Heliotrop (1993), a rotating low-energy building with solar panels that could be continuously directed towards the Sun. Early on, however, he realised that in order to have real impact, solar buildings needed to embark on a transition from eco-niche to mainstream market (Villiger *et al.* 2000). He realised that a first important element of that transition was language. The prevalent terms at the time such as 'passive house', 'low-energy house' or 'energy-saving house' did not seem motivating enough to him, therefore he coined the word 'plus-energy house', referring to a building that would essentially be built according to low-energy or passive house standards, but on top of that generates electricity with solar cells, thus making it a net generator of energy. A second important element was to demonstrate the feasibility of this concept on a larger scale, which also meant paying attention to affordable construction and developing adequate financing concepts. As a major piece of land became available in Freiburg, Germany, after the withdrawal of the French army in 1992, Disch started working on an idea of a large-scale solar housing estate.

5.1.3.3 Initial architectural and business planning (1996–1997)

In the late 1990s Disch further developed his concept and started identifying partners for implementation. He teamed up with Rolf Deyhle, a south German investor who had made a fortune in the real-estate and musical business.[9] The significant financial strength of Deyhle's real-estate company Instag AG convinced the city of Freiburg to offer the land to the consortium.

5.1.3.4 Financial crisis and redimensioning (1998–2000)

In 1998 the project's key investor Rolf Deyhle slipped into financial difficulties following the unsuccessful initial public offering of his musical company Stella AG, and had to withdraw from the planned solar housing estate. Disch had to find new investors and convinced a former chocolate manufacturer and now philanthropic investor, Alfred Ritter, to step in. However, the city of Freiburg, although enjoying a reputation as Germany's solar energy capital, reacted with some scepticism and after a new call for tenders, Disch was awarded less than half of the original land (for a purchase price of €11.6 million), the other half going to a conventional real-estate developer. This was a major setback for the project and made it necessary to re-dimension it from the originally planned 120 units to only 50 units. Disch was also forced to take more financial responsibility than he had originally desired. He found a creative solution, though, by launching a series of four solar real estate funds (Freiburger Solarfonds).

5.1.3.5 Moving into implementation (2000–2005)

After lengthy negotiations with conservative banks, the project was finally able to move towards implementation in the year 2000. In May 2000, in the middle of a crisis of the German real-estate market, the first of the 50 single-family houses, with 85–260 m² living area, were offered for sale. As part of the agreement with the banks, Disch could move into construction of the buildings only as they were sold, which led to a relatively long construction period of six years until the last buildings were completed. In 2003 another important part of the housing estate started, which was a commercial building complex named Sonnenschiff (Solar Ship).[10]

5.1.3.6 Successful completion (2005–2006)

In 2006 the residential housing estate, Solarsiedlung am Schlierberg, and the commercial building Sonnenschiff were completed after a total investment of approximately €40 million (excluding the cost of the land). After much initial scepticism, Rolf Disch and his project are now widely presented as providing a perfect solution to climate change and reducing our dependence on fossil fuels and have won several awards such

---

9 See the Wikipedia entry for *Rolf Deyhle*: de.wikipedia.org/wiki/Rolf_Deyhle (accessed October 2009).

10 See the websites www.sonnenschiff.de (accessed October 2009) and Sonnenschiff-Fonds: www.freiburgersolarfonds.de (accessed October 2009).

as the European Solar Prize[11] in 2002, 3rd prize at the Energy Globe Award[12] in 2003 and the award for Germany's most beautiful housing estate by the Association of the German Paint and Coatings Industry[13] in 2005. A fifth solar real-estate fund is currently being raised specifically for investment in commercial building.

5.1.3.7 Upscaling and future developments (2007 onwards)

Since completion of the project, Rolf Disch continues to develop new ideas. One path to upscaling is a cooperation with a German supplier of prefabricated houses, Weber-Haus, which now offers WeberHaus PlusEnergie,[14] incorporating many of the principles developed by Disch and implemented at Solarsiedlung am Schlierberg. Disch is also involved in promoting solar housing in Africa and Chile. And there are certainly many visitors to the Freiburg solar housing estate that will go home inspired and contribute to the diffusion of sustainable solutions in the building sector.

5.2 Results

5.2.1 Main results

The main result of the project is the creation of one of Europe's largest housing estates with plus-energy houses, including 180 inhabitants and office space for 240 employees. As a financial innovation, the four closed and one currently open solar real-estate funds should also be highlighted. Finally, the case demonstrates that even in basically fruitful environments such as Germany's 'solar capital', Freiburg, ultimately successful projects have to overcome significant barriers of scepticism from public authorities and financial institutions.

5.2.2 Change in sustainability performance

The primary effect of the project is substantially reduced energy use for heating (less than 15 kWh/m²a) and electricity, combined with distributed electricity generation with solar roofs totalling 445 kWp (kilowatt peak) installed capacity and generating approximately 420 MWh per year. For peak heating load in winter, the buildings are connected to a local wood-fired combined heat and power (CHP) plant through a district heating network. Indoor air quality is also improved through controlled ventilation with heat recovery. Other environmental benefits include the use of local forestry

11 www.eurosolar.de/en/index.php?option=com_content&task=blogcategory&id=44&Itemid=24 (accessed September 2010).
12 database.energyglobe.info/ListProjects.aspx (accessed September 2010).
13 www.baunetz.de/meldungen/Meldungen_Preis_Deutschlands_schoenste_Wohnsiedlung_ entschieden_23010.html (accessed September 2010).
14 See www.weberhaus.de/2352.html (accessed October 2009).

products, a ban on PVC (polyvinyl chloride) and the use of natural floor materials. Also, the housing estate has very good public transport connections (bus and tram) thanks to its location close to the car-free Vauban district, therefore there is also an indirect improvement in terms of energy use for transportation.

In terms of social performance, the mix of residential and commercial buildings offers the opportunity for enhanced proximity of work and home, including access to the childcare infrastructure of the neighbouring Vauban district. As the main target group is traditional investors and families, not eco-pioneers, the housing concept is fairly traditional in terms of individual single-family houses, with no particular emphasis on collective space or flexible floor plans.[15] However, for those with more ambition for progressive living styles, once again the neighbouring Vauban district has a lot to offer. Another aspect of social performance is that the project has become an attraction for visitors from around the world, thus providing learning and replication opportunities.

As for economic performance, the key advantage of the project is the relatively low initial cost premium of about 15% over conventional houses. In combination with the low heat demand and the additional income stream for residents through photovoltaic roofs (supported under the German feed-in tariff), this leads to significantly lower life-cycle cost compared with conventional buildings. Hence it can be expected that the project is also going to be a profitable investment for those who invested in one of the five solar real-estate funds (which anticipate an annual return on capital of 4–6%). The absence of fossil fuel consumption creates another important value, which is a hedge against fuel price hikes.

5.2.3 Learning experiences

What started as a visionary idea in the 1990s has become a successful reality 10 years later. Reading reports in the mainstream popular business press in early 2007 (Rees 2007), one would tend to believe that Rolf Disch's solar housing estate in Freiburg should be a role model for all the houses to be built in Germany today. Why is this not the case, and what are the three key lessons learned from the long and partly painful experience of this project?

● Even in seemingly progressive local environments as in the city of Freiburg there is a lot of traditionalism and risk aversion among public authorities and financial institutions. The municipal government's decision to ask a mainstream real-estate development company to build conventional housing on half of the land that was originally planned for the solar housing estate Solarsiedlung am Schlierberg looks narrow-minded in hindsight, but the fear of being accused of incautious decisions at the time seemed to have dominated

● Timing is an important factor influencing success: the launching of a major innovative housing project in the middle of the German real-estate market cri-

---

15 This is why the Solarsiedlung am Schlierberg has only received a 'B' rating at www.oekosiedlungen. de (www.oekosiedlungen.de/schlierberg/matrix-gebaut.pdf [accessed October 2009]).

sis was a tough endeavour. Starting the same project in today's context of low-interest rates for home loans and increased awareness about climate change would probably be much easier

● There is a need for social entrepreneurship. The project would never have come to a successful conclusion without the initiating architect transcending his primary role and taking on significant entrepreneurial risk and demonstrating perseverance over an extended period of time

5.3 Potential for diffusion and scaling up

With the fairly large-scale project in Freiburg now being implemented and serving as a tangible real-life model to demonstrate that plus-energy houses do work and can be built at affordable cost, there is certainly a lot of scope for replication and enhanced diffusion elsewhere. The decision of prefab-house manufacturer WeberHaus to include the plus-energy house in its product line could be an important step towards upscaling, although it remains to be seen whether having this as one of many houses in its catalogue will be sufficient to spark massive consumer demand. Mainstream construction companies and institutional property owners such as pension funds and insurance companies would probably have at least similar leverage. Examples such as SwissRe, which started to introduce office buildings according to the Minergie standard more than six years ago and has now gone a step further to provide incentives for employees[16] if they do the same for their private buildings, seems to be leading in the right direction.

One factor influencing the economics of the project is certainly the generous support for photovoltaics in Germany based on the feed-in tariff of more than €0.40 per kWh. However, given the substantial upfront cost of such roofs, the financial viability of these low-energy houses will not be strongly affected in the first decade of the project's lifetime even without the photovoltaic component. In fact, for some of the houses, ownership of the photovoltaic roof is legally separate from the house below.

Finally, while there is certainly much scope for diffusion of solar housing estates within Europe, the fast-paced North American housing market with its higher levels of energy consumption provides probably even greater potential for improved sustainability performance through a replication of this project.

---

16 See SwissRe's press release about the launch of the 'CO$_{you2}$ reduce and gain' programme: www.swissre.com/corporate_responsibility/coyou2_programme.html (accessed September 2010).

5.4 Overall conclusions

Rolf Disch's Solarsiedlung am Schlierberg (Fig. 5.1) can eventually be assessed as a successful case of sustainable consumption and production. In the 10-year realisation phase, substantial challenges had to be overcome, particularly with regard to finance and regulation, and the re-dimensioning of the project compared with the original plans was a significant setback. However, the initiator of the project has turned into a social entrepreneur, addressing all the financial, regulatory and marketing hurdles on the way. In the context of the now significant evidence of global warming and climate change, followed by even more ambitious targets for energy efficiency, renewable energy and carbon emission reductions set by the German and European governments, this project provides a practical solution that at the same time makes perfect business sense.

FIGURE 5.1 Aerial view of Solarsiedlung am Schlierberg

© Rolf Disch SolarArchitektur

References

CORE (Swiss Federal Energy Research Commission) (2007) *Konzept der Energieforschung des Bundes 2008 bis 2011* (Federal Energy Research Concept 2008 to 2011) (Bern, Federal Office of Energy; www.bfe.admin.ch [accessed October 2009]).

Feist, W. (2006) '15 jähriges Jubiläum für das Passivhaus Darmstadt-Kranichstein: der Faktor 10 ist Realität'; www.passivhaustagung.de/Kran/Passivhaus_Kranichstein.htm (accessed October 2009).

PlusEnergie (2007) *Newsletter* 1/2007 (Solarsiedlung GmbH; www.solarsiedlung.de [accessed October 2009]).

Villiger, A., R. Wüstenhagen and A. Meyer (2000) *Jenseits der Öko-Nische* (*Beyond the Eco Niche*) (Basel: Birkhäuser).

Additional references

Feist, W. (1988) *Forschungsprojekt Passive Häuser; Projektziele: mit einem Kommentar des Autors zur 2. Auflage 1995* (Darmstadt: Institut Wohnen und Umwelt, 1st edn [2nd edn 1995]).

IEA (International Energy Agency) (2006) 'Business Opportunities in Sustainable Housing: A Marketing Guide Based on Houses in 10 Countries'; www.ecbcs.org/docs/Annex_38_IEA_Sustainable_Housing_Marketing_Guide.pdf (accessed September 2010).

—— Solar Heating and Cooling Programme: www.iea-shc.org/task28/index.html (accessed October 2009).

Passivhaus Institut: www.passiv.de (accessed October 2009).

Rohrmann, B. (1994) *Sozialwissenschaftliche Evaluation des Passivhauses in Darmstadt: Passivhaus-Bericht Nr. 11* (Darmstadt, Germany: Institut Wohnen und Umwelt, September 1994).

Wagner, A., and G. Löhnert (2005) '30 Jahre Solares Bauen vom Experimentalbau zum Integralen Gebäudekonzept', in S. Jansen (ed.), *Auf dem Weg in die solare Zukunft* (Munich: Deutsche Gesellschaft für Sonnenenergie).

Wikipedia entry for *Energiestandard*: de.wikipedia.org/wiki/Plusenergiehaus (accessed October 2009).

6

Delivering affordable and sustainable energy: the results of innovative approaches by Woking Borough Council, UK

John P. Thorp
Group Managing Director, Thameswey Ltd, UK

Woking Borough Council in the South-East region of the United Kingdom has long been committed to protecting the environment and it is explicitly stated as one of its top-three priorities. In 1990 it was estimated that Woking produced a million tonnes of carbon dioxide (CO_2) emissions per year. The Council's aim has been to reduce this output to 200,000 tonnes per year by 2090. This reduction is being implemented incrementally in a variety of ways in developments across the town of Woking. Some of these are already in place and will be developed further in the future. This case highlights the political, financial and strategic issues for successful innovation at a community level from the perspectives available in the energy and environmental service company (EESCO), Thameswey Ltd, directly involved in implementing large-scale district combined heat and power (CHP) and renewable energy, primarily with private finance.

Woking is known for its pioneering approach—Woking built what is understood to be the UK's first sustainable energy 200 kWe (kilowatt-electric) fuel cell CHP plant. Woking is estimated to have more than 9% of the total UK installed capacity of photovoltaic (PV) electricity generation and has installed a private-wire-distributed generation system in the Borough, providing electricity, district heating and cooling directly to local customers. Woking's approach to local sustainable energy systems is to supply customers with CHP on private wire and/or on 'virtual' private energy networks as well as implementing energy and environmental services in both the public and the private sectors. The key to the Council's success is the combination of technical innovation (such as CHP, absorption cooling, private-wire systems, etc.), partnership with the private sector, financial and commercial innovation and the use of a local electricity balancing and trading system.

Woking Borough Council has taken a political lead and developed its own Climate Change Strategy for the Borough, taking the view this is not something to be ignored—we all have to take action now.

The recently updated Climate Change Strategy for Woking (WBC 2008) aims to build on the Council's previous environmental successes notably in energy efficiency and to take a carbon-neutral approach to the future of services and activities within the Borough. The strategy sets out a range of options that aim to reduce CO_2-equivalent emissions (mitigation) and take further measures to enable the habitats within Woking to evolve in response to climate change (adaptation).

The Strategy builds on an original Energy Efficiency Policy. Woking has been introducing a range of innovative measures to protect the environment and reduce pollution since 1990; adopting numerous energy and water saving techniques for Council-owned buildings and public places. The revenue saved has been reinvested in other environmental measures to further improve energy efficiency. Woking's Strategy is a more comprehensive step in taking forward environmental good practice at the local level; the Council is taking action to reverse the build up of greenhouse gases and is meeting and exceeding national targets, such as the targets of the UK government's Royal Commission on Environmental Pollution to reduce CO_2-equivalent emissions by 60% by 2050 and by 80% by 2100.

6.1 Case Description

6.1.1 Overview

Woking Borough Council is tackling climate change through a number of key areas:

- Planning and regulation
- Energy services
- Waste
- Transport
- Businesses
- Community
- Procurement
- Education and promotion

- Management of natural habitats

- Adaptation to climate change

The Woking Borough Council Climate Change Strategy (WBC 2008) covers the whole spectrum of the Borough's energy use and climate change emissions: from power, heat, water, waste disposal and transport. The Strategy includes proposals for recycling and reducing waste and recovering energy from non-recyclables through, potentially, anaerobic digestion, gasification and CHP.

The principles of the strategy incorporate three overarching themes:

- Reduction of CO_2-equivalent emissions

- Adaptation to climate change

- Promotion of sustainable development

By encompassing a wide range of different objectives, spanning the breadth of services and activities undertaken by the Council, the Strategy ensures that the objectives for responding to climate change are embraced holistically across the Council. Although its context is long-term, the Strategy includes targets and actions to be achieved in both the short and the medium term. It aims to be a flexible document and is reviewed regularly. The Strategy document is a crucial tool in formalising the ambitions of local stakeholders, for example through Climate Neutral Development Good Practice Guide, promoting voluntary cooperation between the parties involved in the local development process to work towards achieving an 80% reduction in CO_2-equivalent emissions. By looking at the everyday needs and services required by the public and provided by Woking Borough Council (such as waste management, housing, etc.) the Council envisages that local residents and businesses can all work together to cut down the causes of climate change and in the process make the living environment within the Borough more sustainable.

One of the principal ideas of Woking Borough Council in connection with development and land use is the promotion of the concept of an 'environmental footprint'. This refers to the CO_2-equivalent emissions that land use, be it a field, office block or housing estate, produces from heating, methane generation and so on. The aim is to encourage a lower, less harmful level of CO_2 in the environmental footprint of a site, with the overall objective that any new land use reduces CO_2 emissions by 80% compared with the previous use. This would mean that if an office block were to be replaced with a housing estate, the housing estate would have to incorporate sustainable and renewable energy measures to result in significantly lower CO_2 emissions than those produced by the office block.

6.1.2 Context: social factors

Achievement of an integrated and sustainable community is the major challenge for Woking Borough Council. Despite the position of the town in one of the more affluent regions of the UK the level of disadvantage within Woking's community is significant, whether this is based on access to affordable housing, ethnicity, economic vitality, education or health. These all combine to mean that for some living or working in the community the quality of life is poor. For the disadvantaged and vulnerable not only is the quality of life sometimes poor but also the risk of premature death is high. For the majority of those living and working within the Borough the quality of life is good and the risk of premature death is low. It is in trying to raise living standards in an affordable and environmentally responsible way that Woking faces a significant challenge and where the need for community leadership is essential.

6.1.3 Actors and their roles and perspectives

6.1.3.1 Primary actors

The elected leader of Woking Borough Council's Executive, the key decision-making body within the Council, demonstrates an ongoing commitment to championing the interests of sustainability in the Climate Change Working Group, tasked with taking forward the objectives of the Council's Climate Change Strategy through a rolling action plan.

The Council's Executive is made up of seven councillors and meets twelve times a year. In May of each year, the Council elects the Leader of the Executive and appoints six councillors to the Executive. The Executive is accountable for decision-making within the Council's overall policies and budget framework and recommends to the Council new policies and decisions outside its remit, ensuring the consistent application of corporate standards.

When major decisions are to be discussed or made, these are published in the Executive's forward plan of key decisions. This plan lists all the decisions to be taken by the Executive, that is, decisions likely to result in the Council incurring expenditure that is significant, or making savings that are significant, having regard to the Council's budget, or significant in terms of its effects on communities living or working in an area comprising two or more wards or electoral divisions in the area of the Council. All meetings are open to the public, except on rare occasions.

Local government is a complex and diverse industry. Local councils in the UK are under constant financial pressure to move forward an ever-increasing agenda of improvements, both internally generated and imposed by central government. This is against a backdrop of political pressure to cap or reduce local taxes and decreasing financial support from central government. Historically, councils have concentrated on a one-year planning cycle, which inhibits their ability to deliver well-structured developments. Longer-term strategic planning cycles, particularly with regard to financial management, are the desired route. It is exactly for this reason that Woking councillors decided to protect environmental initiatives from the usual budgeting processes. This required a suitable financial and delivery vehicle, and Thameswey Ltd was incorporated.

The Chief Executive of Woking Borough Council, not an elected individual, is expected to make a difference in personally moving this agenda forward across the organisation. Through the Chief Executive championing equality and climate change together and the Council working in formal collaboration with a significant number of commercial partners and community leaders, progress has been made to underpin the long-term sustainability of Woking's community.

When contemplating the effects of climate change on the local area, it is important to recognise the need for the community to work together. Woking Borough Council has worked with partners, including local people, through focus groups (Woking's Citizen Panel; representatives of the Borough's population) and a Climate Change Strategy Group (with members drawn from Council staff, politicians, businesses and residents) to develop a Community Strategy for the Borough with aims such as a clean, healthy and safe environment at its core. Not only did the groups help formulate the branding of the Strategy but also they were useful in informing residents of the Strategy and of the Borough's initiatives (Curran 2007).

The Climate Change Strategy has helped contribute to the Community Strategy by, for instance, using alternatives to fossil fuels. This is not only cutting greenhouse gas emissions but also improving air quality locally and helping to contribute to a clean, healthy, affordable and safe environment.

6.1.3.2 Secondary actors

Woking Borough Council has established a role as a strong community leader that sets the very highest standards in energy efficiency and tackling climate change. However, the Council has sought to 'take the community along with it' and believes community engagement to be vital, to ensure its example is followed by others. When developing the Climate Change Strategy, independent market research consultants were commissioned to hold focus groups to ascertain how the public felt about the issue (Fig. 6.1). This revealed that despite high levels of environmental awareness claimed by many of the respondents, understanding of climate change was still very limited and some were sceptical about the causes of climate change. However, there was good support for and recognition of the actions the Council had taken. This research demonstrates the Coun-

FIGURE 6.1 Woking focus group

Photo: J. Thorp. © ECSC Ltd

cil's commitment to engage the local community in its decision-making and to seek to act in an open and transparent manner in respect of an issue that can be controversial. The community with which Woking Borough Council engages includes:

- The public (who elect 33% of Council members every year)
- Commercial partners
  - Thameswey Ltd and its energy and environment related subsidiaries such as Thameswey Energy Ltd
  - Sponsors, such as those for the Woking Park fuel cell CHP
  - BP Energy Ltd for PV schemes
  - Xergi A/S for CHP schemes
  - English Partnerships for embedded generation projects in communities outside Woking
  - Woking Borough Homes Ltd for small CHP schemes
- Government agencies, such as the Department for Environment, Food and Rural Affairs (Defra), the Environment Agency, the Carbon Trust and the Energy Saving Trust
- Local Governments for Sustainability (ICLEI)
- Business organisations such as chambers of commerce

The good practice guidance for planners and developers promoting climate-neutral development was developed in specific partnership with the preferred housing associations, the Home Builders Federation, Three Rivers Water, the Environment Agency, South-East England Development Agency, the Construction Industry Research and Information Association (CIRIA) and the Carbon Trust (government agency). It was formally endorsed by the UK Climate Impacts Programme (UKCIP) and the South-East Climate Change Partnership.

6.1.4 Case history and development

Woking, through the interests of its councillors, and reflecting residents' concerns, has long been committed to protecting the environment. This commitment to tackling climate change has been widely recognised by the government through its Beacon award scheme, which promotes excellence in delivery of local authorities' services to their communities. In 2001 Woking received the Queen's Award for Enterprise in recognition of its approach to sustainable community energy systems.

The Council has been awarded Beacon Council status for Sustainable Energy (2005/06) (IDEA 2005), Promoting Sustainable Communities through the Planning Process (2007/08) (IDEA 2007) and for Tackling Climate Change (2008/09) (IDEA 2008).

The initial Climate Change Strategy was first adopted in 2002 to define specific, measurable local objectives, with a clear action plan and monitoring system, including annual assessments of achievements against targets. The Strategy continues to acknowledge the challenges that the community must face if national targets for CO_2 emission reductions are to be achieved and translates these into its own challenging objectives, in step with realising national targets.

The reduction in CO_2 is being implemented incrementally, in a variety of ways, in developments across the town of Woking, combining technical, policy, service and delivery innovations (see Table 6.1).

TABLE 6.1 **Technical, policy and service and delivery innovations introduced by Woking Borough Council, by year**

Year	Innovation
Technical innovations:	
1990	Energy efficiency strategy implementation
1996	Installation of small-scale community combined heat and power (CHP)
2001	A private-wire-distributed generation system providing electricity, district heating and cooling directly to local customers
2001	Woking Park Fuel Cell CHP: the UK's first 200 kWe fuel cell CHP
2004	Absorption cooling
2005	Extension of town-centre CHP
2005	Photovoltaic installations (ongoing)
2007	Implementation of a range of micro-generation technologies, with Energy Centre for Sustainable Communities Ltd (ECSC) and Energy Saving Trust (EST) guidance on best practice in a pilot low or zero carbon house
Policy innovations:	
2004	Introduction of Climate Neutral Development Guidance
Service and delivery innovations:	
1999	Establishment of an energy and environmental service company called Thameswey Ltd (see Section 6.1.4.1)
2001	Local electricity balancing and trading system
2008	Establishing a home energy improvement scheme

Each area for action under the Climate Change Strategy has a series of targets and actions with an associated time-scale of 1–3 years, 3–5 years or 5–10 years. Progress within actions is monitored and reported to the Climate Change Working Group on a quarterly basis. Annual reports incorporating information on energy efficiency and CO_2-equivalent emissions savings are produced.

6.1.4.1 Establishing an energy and environmental service company: Thameswey Ltd

The overarching strategy requirement for Woking was to facilitate the achievement and implementation of the Council-led sustainability policies within the Borough. Having had significant success with small-scale schemes within the Borough, the Council sought to create a new business with a specific organisational format, an energy and environmental service company (EESCO). This was for several reasons: to allow commercial projects to be run both inside and outside the Borough, to give enough autonomy from the election cycle to respond effectively to opportunities, to offer the capacity of autonomous budget management over the longer term, to harness external investment and to be profitable while remaining responsive to the Council's political objectives.

The EESCO Thameswey Ltd and its energy and environment related subsidiaries (the Thameswey Group) were established by the Council through cross-party collaboration and agreements to assist the Council further its objectives across its Climate Change Strategy and Community and Housing Strategy without giving rise to an increase in council tax.

The initial capital for Thameswey Ltd came through reserving funds from the energy savings within Council properties. Crucially, all profits from the operations of the Thameswey Group are utilised for the furtherance of energy efficiency and sustainability investment within the Borough. The financial consequences for the Council's budget in progressing the sustainability agenda are therefore minimised

The EESCO Thameswey Ltd has successfully developed, through offerings to the market, solutions to the technical, political and managerial issues inherent in the implementation of long-term energy projects. To date Thameswey Ltd, through its energy and environmental activities, has been successful in progressing initiatives that have enabled the Council to achieve widespread recognition for its work on climate change.

Thameswey Ltd promotes energy efficiency, energy conservation and environmental objectives by providing energy and/or environmental services:

- Developing and implementing new technologies for the production and supply of energy

- Producing and supplying energy (and any related by-products) in all its forms

- Acquiring and holding interests in the share or loan capital of any company or corporation and, in particular, companies engaged in energy and/or the environment business

- Providing financial, managerial and administrative advice services and assistance

- Making facilities and services available for its customers and customers of companies in which it holds an interest

In May 2000 Thameswey Ltd set up an unregulated public–private joint venture energy services company called Thameswey Energy Ltd (TEL) to finance the first energy station in Woking town centre. Crucial to the setting up of this company was the recruitment of a large financial investment organisation as a shareholder. This was initially a Danish pension fund, although later company changes resulted in Thameswey Ltd developing a strong business relationship with a UK investment bank, which became the major shareholder.

TEL also brings together the local authority with a Danish CHP design-and-build company (Xergi A/S) as a partner. TEL operates the Council's installations in Woking Park that comprise a hydrogen fuel cell, PV cells and CHP. These provide heat and power for Woking's leisure centre, swimming pools and lighting in the park.

TEL has continued to deliver a range of sustainable and renewable energy projects in order to meet the Council's Climate Strategy ambitions. TEL acts as a contractor to Woking Borough Council to invest in CHP plant (energy stations), to sell heat and power in an environmentally friendly way, with a view to improving the environment within the Borough.

TEL can provide local sustainable energy services to other local authorities, public bodies and the private sector both within and outside Woking. TEL is the division of the Group that furthers the Council's energy objectives, both in the Borough and elsewhere. For projects outside of the Borough of Woking the company policy is to establish a subsidiary so that there is a clear division between activity for the direct benefit of Woking and activity that provides an indirect benefit. Currently, TEL has two operational subsidiaries, Thameswey Central Milton Keynes Ltd and Xergi Services Ltd (who provide operations and maintenance support for all installations). The TEL financial forecast for the period to 2030 has been developed with a target return for shareholders of 8%, in accordance with the strategy approved by the Council. TEL is progressing the following:

- Implementation of CHP projects within the Borough of Woking on the basis of a target return for shareholders in accordance with the financial arrangements approved by the Council in December 2004

- PV installations on Council property through a partnering arrangement with BP Solar Ltd (manufacturer, supplier and installer of the PV panels), wherever the financial case can be supported and financial assistance from other agencies can be obtained with a view to achieving an installed capacity of 1 MWpe

- Small-scale or medium-scale wind turbines

- Assistance to the Council towards achieving its objective of the conversion of existing housing in the private sector to achieve 1,000 low-carbon homes with embedded micro-generation

- Pursuit of projects outside the administrative area of Woking with a target return for shareholders of 12%

TEL provides customers with sustainable energy at an affordable cost in comparison with conventional energy. It is able to do this, despite the higher cost of the sustainable energy plant, because of the payback from the plant through combining the revenue from selling heating, cooling and electricity.

In June 2007 Thameswey Ltd purchased the energy efficiency and climate change business, the Energy Centre for Sustainable Communities Ltd (ECSC). ECSC, in addition to carrying out commercial consultancy on climate change and planning and development issues throughout the UK and Europe, also provides all internal professional staffing resources for the Thameswey Group.

6.1.4.2 Woking Borough Council energy projects

In Woking the combination of long-term political and financial support structures has enabled the barriers of short-term planning and low or high priced working capital to be crossed. The case studies below emphasise the absolute requirement for innovative thinking, for all variables to be understood and for risk to be minimised and utility maximised.

6.1.4.2.1 Woking Park fuel cell combined heat and power

Woking Park fuel cell CHP (Fig. 6.2) was developed in partnership with the Council and TEL. The system was sponsored by the UK Department of Trade and Industry, Advantica Technologies Ltd,[1] US Army Corps of Engineers Research and Development Center (ERDC) (US Department of Defense),[2] The National Energy Technology Laboratory (US Department of Energy)[3] and Woking Borough Council. The associated community energy systems were partly sponsored by Defra. Contractors included UTC Fuel Cells LLC[4] and BTU (Heating) Ltd.[5] As a demonstration facility full monitoring has been carried out throughout. UTC has now been developing a 400 kWh unit and it is expected that negotiations will commence to retrofit the upgraded technology.

FIGURE 6.2 Woking Park fuel cell combined heat and power: public art

Photo: J. Thorp

1 www.advanticatech.com
2 www.erdc.usace.army.mil
3 www.netl.doe.gov
4 www.utcpower.com
5 www.btu-heating.com

The fuel cell is designed to support the swimming pool in the Park's heating and power systems and as well as the Park's lighting. Excess heat produced is used to power the centre's air conditioning, cooling and dehumidification requirements via heat-fired absorption cooling.

The CHP station is also designed to provide energy services for the Leisure Centre, with surplus electricity exported to the Council's sheltered housing schemes. For energy flows, see Figure 6.3.

6.1.4.2.2 Addressing fuel poverty

As part of the Council's developing strategy to address fuel poverty, the Council has used three approaches, each unlikely to have developed through market mechanisms. First, the Council provides energy services (electricity and heating) to its sheltered housing tenants for a known weekly charge included in the rent by use of private-wire electricity distribution systems, CHP and renewable energy systems. This means that tenants spend less than 10% of the state pension income on heating and electricity.

Second, the Council recognises that a proportion of the population remains 'out of reach' because it does not approach agencies that can give it advice, and many people living in households in fuel poverty are not receiving help. The Council entered into a contract with ECSC to utilise The Energycare Network (TECN) to provide energy efficiency training for people working in the community who are known and trusted by the clients, such as home carers, Citizens Advice Bureau (CAB) welfare rights workers, health workers and local people who could become energy champions. The training covers how to identify a householder experiencing the effects of fuel poverty, the use of TECN referral process and how to give effective advice. The training sessions highlighted the links to improved health from the improvement of domestic energy efficiency. Three workshops, called 'Engaging ethnic minorities and the hard to reach', were held. TECN will continue to build on the 20 front-line professionals and voluntary sector workers who have already undergone training in Woking.

Third, the Council also set up a pilot project to increase the uptake of TECN, Homelink and Woking Borough Council services within the Black Minority Ethnic (BME) communities in Woking. Following feedback collected after a TECN event, and a meeting with the secretary of the local mosque, it was found that the best way to engage the local ethnic community was by word of mouth rather than inviting large groups to a formal presentation. It was decided that an energy-efficient 'show house' would enable members of the local community to see and hear for themselves the types of measures and assistance that was available to improve their homes, passing on their learning by word of mouth.

6.1.4.2.3 Guidance for planners and developers

Woking Borough Council has developed good practice guidance for planners and developers promoting climate-neutral development. Since 2006 this has been implemented to encourage voluntary cooperation between parties involved in the local development process with a view to achieving an 80% reduction in CO_2-equivalent emissions and adapting to climate change. The guidance comprises a range of climate change mitiga-

FIGURE 6.3 Woking Park fuel cell combined heat and power: energy flows

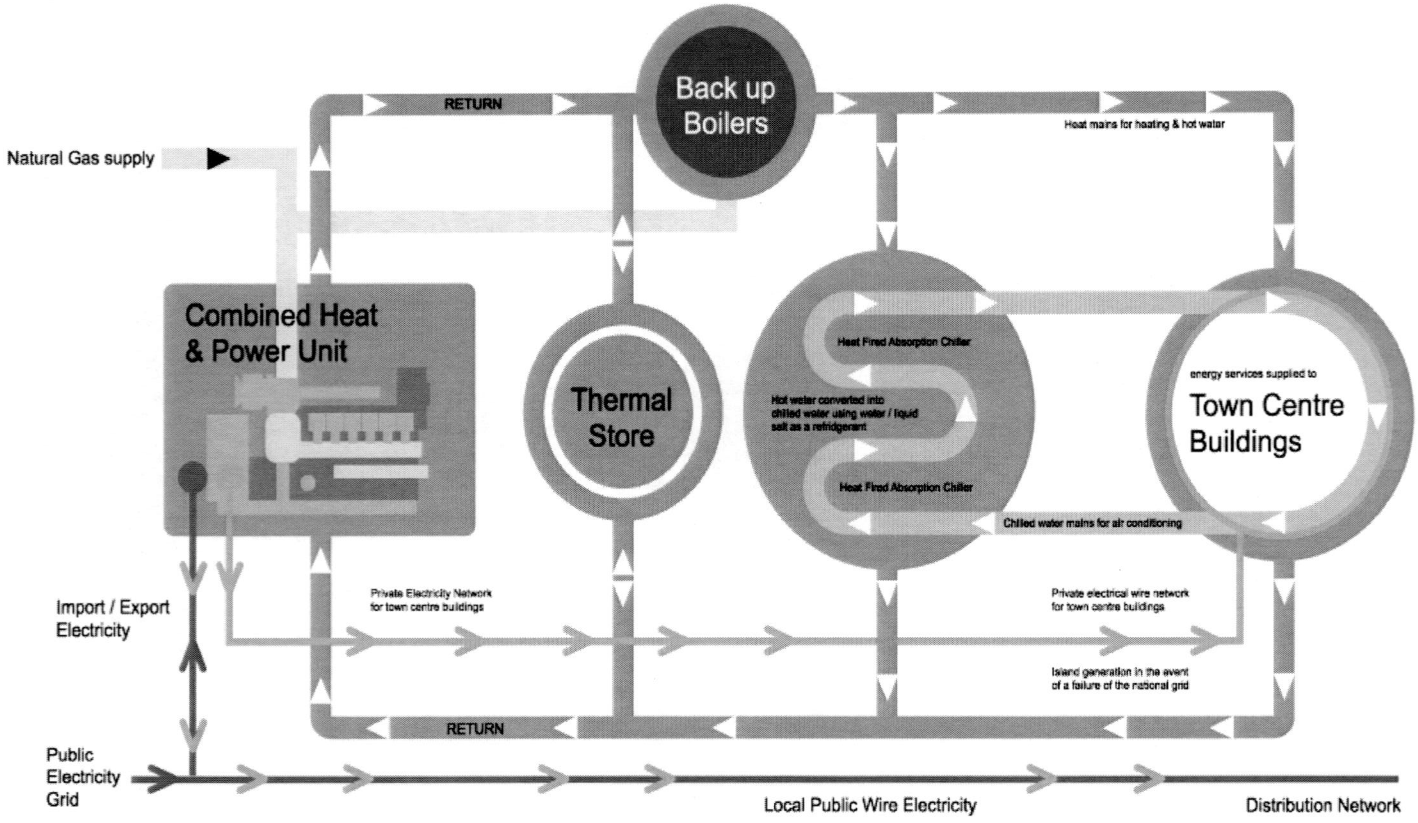

Source: J. Thorp. © Thameswey Ltd

tion and adaptation measures, including sustainable energy. With the Council as the catalyst it was developed in partnership with a group of key stakeholders, including preferred housing associations, the Home Builders Federation, Three Rivers Water, the Environment Agency, South-East England Development Agency, CIRIA and The Carbon Trust. As this is a voluntary code, the whole area needed to be demystified and the cost implications and benefits clearly set out. The guidance includes case studies of recent development schemes in the Borough, with cost and CO_2 emissions comparisons between grid energy and sustainable energy. Woking's Climate Neutral Development Guidance has been formally endorsed by UKCIP and the South-East Climate Change Partnership.

Through the work of Woking and others, as Beacon councils, many other Boroughs are now taking a view that planning is an effective tool for local authorities to use in taking forward the vision for their areas as set out in their community strategies. The planning process offers opportunities to influence how areas develop. More effective community involvement is best achieved where there is early engagement of all the stakeholders in the process of plan making and bringing forward development proposals. This helps to identify sustainability issues and problems at an early stage and allows dialogue and discussion of the options to take place before proposals are too far advanced.

6.1.4.2.4 Wind energy

Wind energy is considered to have the potential to make a significant contribution to the Council's Climate Change Strategy through the production of electricity from a renewable source. Surrey has not been identified as a prime location for large-scale wind farms because the potential for wind energy in other parts of the UK is significantly greater. Major commercial operators providing energy for the wholesale market will target those areas of maximum wind energy generation, and there remains significant under-utilisation of this source of energy across the UK. However, wind energy is available in Surrey. Based on the Department of Trade and Industry (DTI) database of background wind speeds, the wind speeds in Woking are 4.6 metres per second (m/s) at 10 m above ground level, 5.3 m/s at 25 m above ground level and 5.8 m/s at 45 m above ground level.

Large wind turbines, usually some 45–70 metres above the ground with 25 m plus rotor blades, will operate with wind speeds from about 4 m/s; efficiency improves up to about 17 m/s. Small, domestic-scale, wind turbines, usually 8–15 m high with 1–3 m rotor blades, typically operate with wind speeds from 5.4 m/s. As the small turbines are lower than large wind turbines they are likely to operate less frequently and less efficiently than the larger turbines.

Having considered this the Climate Change Working Group supported the proposal that the Council, through its Thameswey operations, should progress the installation of large-scale wind turbines in a phased programme, with an initial phase of at least three turbines with individual generating capacity of some 1.3–1.5 MWe, together generating some 3.9–4.5 MWe.

Large-scale wind turbines are considered economic, particularly if the energy produced can be sold directly to customers rather than to the wholesale market. The unique character of TEL means that it can secure retail value from energy sales and claim the renewable obligations credits. It was therefore proposed that, as part of the TEL business plan for energy production within the Borough of Woking, it should progress the installation of three large-scale wind turbines.

Anticipating community consultation through the planning application process, the Council commissioned ECSC to carry out research among the local community on opinions and perceptions relating to wind turbines in order to inform the Council's policy development on wind energy. To date, this initiative, although still being contemplated, has been re-scheduled within the business plan. It is proposed to bring forward, when the opportunities arise, small to medium scale wind turbines, that is, turbines that are larger than domestic turbines but suitable for locations such as small recreation grounds.

6.1.4.2.5 Micro-generation for 1,000 low-carbon homes

The Climate Change Working Group identified the government's initiatives on microgeneration as a further opportunity to progress the Council's Climate Change Strategy. Half the existing UK stock of 23 million homes is more than half a century old, and over 80% of the building stock for the year 2025 has already been built. Yet, from a policy perspective, effectively tackling the problem of existing buildings is difficult. Setting out targets and a framework for low- or zero-carbon buildings is comparatively straightforward—tighter building and planning regulations—but solutions for upgrading late nineteenth or early twentieth century houses can be found.

The Low Carbon Homes Programme continues Woking's leadership in practical carbon reduction and tackles one of the most difficult sources of greenhouse gases, existing privately owned homes. The concept of private homes being 'micro' generating sites with net export of electricity and low or zero CO_2 emissions offers an exciting opportunity to raise the profile of the role of residents in achieving a local sustainable community.

Initial informal discussions with the Energy Saving Trust (EST) indicated an interest in this initiative, which would fall within its Low Carbon Buildings Programme. To move this forward an existing 1930s empty property (Oak Tree House) has been converted to raise its energy efficiency standard and to implement a range of technologies:

- Insulation (walls, ceiling, windows)
- Energy-efficient lighting and fixtures
- Energy-efficient appliances
  - Washing machine
  - Refrigerator
  - Dishwasher
  - Standby (current) control switches
- Rainwater harvesting-to-toilet flush systems

- Water-efficient fittings
  - Aerating shower head
  - Dual-flush toilet
  - Tap flow regulation valves
  - Water butts and various other smaller measures

- High-efficiency condensing gas boiler

- Micro-generation systems; solar PV cells, solar hot water and a biomass boiler

Other small but important elements of installation and specification include: pipe lagging, minimised plumbing distance to points of use, small-bore pipes, draught blocking and so on.

The house has been transferred from the Council to Thameswey Energy Ltd, to upgrade it with the help of partners and to use it as the base for the project. The Council has the intention of later putting it into a Thameswey subsidiary, Woking Borough Homes Ltd, to rent at an intermediate market rent, when the first phase of the Brookwood Farm is completed.

The Low Carbon Homes project will work within the strategic framework of the Climate Change Strategy towards a target of carbon neutrality. The programme will aim to recruit occupants of privately owned homes within Woking to reduce their carbon emissions as low as practically possible, ultimately heading towards carbon neutrality and an annual reduction of approximately 5,800 tonnes of CO_2. Practical water conservation measures will form a secondary target, supporting the first phase of homes to reach a 20% reduction in water consumption, saving approximately 12,410 million litres of potable water every year.

The first phase of Low Carbon Homes will recruit 1,000 homes by March 2012, focusing on minimising the CO_2 from standing energy in the homes of participants. Owing to the strong connection between energy and water, a 20% reduction of mains water consumption is proposed as a secondary target for programme participants. There will be a staggered recruitment of 100 residents in the first year, 200 in the second and 350 in the final two years of initial implementation. This aims to reflect an increased momentum as the programme progresses

Low Carbon Homes will address a number of the barriers to the take up of energy and water efficiency and micro-generation and encourage residents to work towards carbon neutrality but will initially focus on standing energy and water. Measures that will be favoured and actively promoted by the programme will be:

- Simple—robust and reliable measures that do not involve complex installation or maintenance

- Practical—measures which can retrofit into existing Woking homes that deliver CO_2 and water reductions

- Effective—proven technologies that can deliver the best measurable results

Low Carbon Homes proposes to do this through:

- A menu of energy and water retrofits

- The Oak Tree House—an 'eco-house' display home

- Structured financial products for capital investment and repayment

- Micro-generation systems available to residents through a long-term lease agreement

- Residents being offered a purchase–lease arrangement to purchase the system over the course of an agreed time-frame through a contractual arrangement

- An energy export and supply contract to offer additional incentives to residents

- UK government grants through the Low Carbon Buildings Programme (most of the example measures installed at Oak Tree House are eligible for these grants)

- A retail premises or 'eco-shop'

- Links and networks with existing projects, services and organisations

An analysis of the impact of tariff rates demonstrates the importance of both energy efficiency and of tariff structure. For example, a resident installing a 1 kWp solar PV system, without implementing energy efficiency and receiving only £0.05 per kWh produced, would not recover the costs of that installation for nearly 80 years. However, a matched feed-in tariff of £0.10 per kWh halves this to around 40 years and, combined with a 30% reduction in energy consumption, can pay for the solar PV array in an attractive 16 years: a 6.2% annual return.

All Woking residents have access to the same independent information; however, participants of Low Carbon Homes will receive additional communications to support them to implement actions at home. These will include:

- Low Carbon Homes newsletter

- Telephone support

- Invitations to events, such as workshops at Oak Tree House

- Feedback and evaluation

6.1.4.2.6 Energy efficiency schemes

Thameswey Ltd has also been the vehicle for delivering the Council's home energy improvement schemes under the government's Home Energy Conservation Act 1995 (HECA) objectives in partnership with others. In 2006, 2007 and 2008 almost 5,000 private homes with elderly residents have benefited from free insulation of roof spaces and cavity walls. Woking has been able to secure additional funds for the scheme from Energy Utilities (under the Energy Efficiency Commitment) to help finance the insulation for the very elderly and/or disabled on benefits, through the government's Warm Front and Energy Efficiency Commitment Schemes.

6.1.4.2.7 New and extended combined heat and power capacity

Wherever practical Thameswey Ltd will pursue new CHP opportunities in the Borough, within the approved guidelines set out by the Council. Small schemes are already being implemented in association with Woking Borough Homes Ltd together with various schemes under the Community Energy Programme investment by the Council in its existing housing stock, with the assistance of financial support from Defra.

In 2006 the town centre network was extended and the existing cooling systems were upgraded to provide additional capacity for both electrical generation and cooling. Thameswey Ltd has implemented the extension of the town centre CHP system to include the Lightbox Museum and Arts Centre (see Fig. 6.7).

6.1.4.2.8 Photovoltaic installations

Thameswey Ltd has also implemented PV installations where the economics permit. PV is an expensive installation but it has long-term benefits and low operating costs. The company has assisted in the procurement of PV for the Albion Square canopy, with support of a 50% grant from Defra. The net cost to the Council for the canopy as a result of this joint work is the same as if it had installed only conventional glass (Fig. 6.4).

FIGURE 6.4 Photovoltaic roof: the Albion Square Canopy

Photo: J. Thorp

With the canopy, the Borough have some 600 kWpe of PV installations. Thameswey Ltd will continue to explore opportunities for increasing this to the target level of 1 MWpe on Council corporate assets.

6.1.4.3 Thameswey Central Milton Keynes Ltd

Thameswey Central Milton Keynes Ltd (TCMK) has been established as a Thameswey Ltd subsidiary company to provide CHP energy services to the central area of Milton Keynes, complementing the activities in Woking. The land in Milton Keynes is owned by English Partnership (Commission for New Towns), which has entered into a long-term contract with TCMK. The initial CHP phase of the project is for an investment of some £8.3 million. The 25-year business strategy is underpinned by the objective of achieving a 12% return for shareholders.

The initial CHP phase comprises a number of different development blocks and developers. The development blocks nearest to the energy station have contracted with TCMK to take supplies (with developers Crest Nicholson,[6] Frontier,[7] Abbeygate and Hampton Brook[8]). English Partnerships (now Homes and Communities Agency) has initiated further blocks for which TCMK is negotiating supply. It is envisaged that the whole development will take some seven to ten years and could require two further CHP stations, bringing the total investment to some £40 million.

The arrangements with English Partnerships provide for the ongoing extension of the project through Central Milton Keynes with a minimum target return of 12%. Should English Partnerships require a CHP supply to be made to part of the area it can call on TCMK to do so. If TCMK can establish it can achieve a 12% return (or more) it is obliged to proceed with the phase, if TCMK cannot achieve the target return of 12%, then it is obliged to proceed only if English Partnerships meet the financial shortfall.

6.2 Results

6.2.1 Main results

Woking Borough Council has sought to provide affordable, sustainable energy for residents as an integral part of their Climate Change Strategy. The Climate Change Strategy defines specific, measurable local objectives with a clear action plan and monitoring system in place and annual assessment of achievements against targets. The Strategy acknowledges the challenges that the community must face if national targets for CO_2 emission reductions are to be achieved and translates these into its own challenging objectives.

Simultaneously, Woking has sought to tackle fuel poverty by reducing heating costs for Council accommodation to a percentage of the state pension and has pursued energy conservation. Of the 36,941 households in Woking, over 12,000 have so far taken advantage of the Council's energy conservation schemes. The Council has helped almost 5,000 private-sector households with energy conservation grants since 1996/97.

6   www.crestnicholson.com
7   www.frontier-estates.com
8   www.hamptonbrook.com

FIGURE 6.5 Photovoltaic installation at Sunnyside Residential

Photo: J. Thorp. © Woking Borough Council

FIGURE 6.6 Photovoltaic cell installations at Brockhill Sheltered Housing

Photo: J. Thorp. © Woking Borough Council

Woking's approach to local sustainable energy systems is to supply customers on private-wire CHP and/or renewable energy networks as well as implementing energy and environmental services in the public and the private sectors. The Council operates 57 community heating sites, including 3 PV sites (for example, see Fig. 6.5).

Thameswey Energy Ltd operates six private-wire CHP sites on behalf of the Council, of which five have PV installations. These sites include an innovative mix of technologies combining small-scale CHP and PV installations to deliver a 100% renewable or sustainable energy source providing affordable heat and electricity directly to residents in sheltered housing (see Fig. 6.6).

In addition to these technologies the Council has completed a rainwater harvesting system in Woking Park that will reduce water pump energy consumption and has installed off-grid PV and wind energy lighting systems in three locations in the Borough.

Woking's Climate Neutral Development Guidance is believed to be the first of its kind in the UK and promotes voluntary cooperation between parties involved in the local development process with a view to achieving an 80% reduction in CO_2-equivalent emissions and to mitigating against the local effects of climate change.

FIGURE 6.7 Woking Plans

Bordered circles indicate areas identified for possible future development of district heating networks and private wire.

Source: J. Thorp (2007). © Thameswey Ltd

Woking Borough Council maintains share capital investments in Thameswey Ltd and its subsidiaries, when the projects are brought forward, meet the criteria set out by the Council (8% internal rate of return on shareholder investments). This means that Thameswey Ltd continues to operate its business with a view to break even over the long-term project cycle, having invested the returns from its activities to further the environmental and housing objectives of the Council within the Borough.

Woking has been a leader in the UK in the development of local sustainable community energy, for which they received the Queen's Award for Enterprise: Sustainable Development 2001, a Beacon Council award for Environmental Sustainability in 2004 and a Beacon Council award for Planning in 2007. Woking is the only local authority ever to receive a Queen's Award for Enterprise, recognising the Council's unique achievement in sustainable energy. For residents, the greatest benefit has been that Woking has achieved these objectives in a way that has been self-funding, working with commercial and non-commercial partners to meet environmental goals. Woking also demonstrates excellent practice in the field of partnership—many of its achievements are possible only through inter-agency working.

6.2.2 Change in sustainability performance

Table 6.2 presents a comparison of Woking Borough Council energy performance data for 2007 with a baseline year of 1990.

TABLE 6.2 **Woking Borough Council energy performance data: a comparison of data for 1990 with those for 2007**

Measure	Change
Corporate figures:	
Energy consumption	−28%
Carbon dioxide emissions	−55%
Sustainable energy self-generation	+60%
Renewable energy self-generation	+2%
Borough-wide:	
Energy efficiency of residential property	+35%
Carbon dioxide emissions	−21% (2008)
Number of households assisted with energy conservation grants	5,072[a]

[a] Absolute number, not a percentage change.

Some 98% of Council dwellings can be heated to a comfortable temperature for 10% of income or less (for sheltered housing) or £10 a week or less (for non-sheltered housing). The Council has secured a UK government (Defra) grant to provide CHP and affordable heat and private-wire electricity to an additional 240 sheltered housing dwellings.

6.2.3 Learning experiences

A key factor in the Council's success is the combination of technical innovation (such as CHP, absorption cooling, private-wire systems, etc.), partnership with the private sector, financial and commercial innovation and the use of a local electricity balancing and trading system. Although Woking had been successful in implementing small-scale local community energy systems, to capitalise fully on its sustainable energy innovation it needed the finance and expertise of the private sector to finance and implement large-scale projects. The formation of Thameswey Ltd and Thameswey Energy Ltd in 1999 enabled the Council to increase its distributed generation capacity.

Woking's achievements in sustainable local energy owe much to the political leadership of the Council in championing the issues of energy efficiency and sustainability. The active participation of the Council's political leaders in promoting sustainability has been one of the main contributory factors in achieving mainstreaming of sustainability throughout the Council's business activities.

Woking have partly overcome the regulatory barriers to sustainable energy, namely, the high costs and levies incurred on electricity bills under the New Electricity Trading Arrangements, by taking advantage of the Electricity (Class Exemptions from the Requirement for a Licence) Order 2001. However, the exempt licensing regime limits exempt supply capacity to domestic customers and limits exports over public wires. Woking continues to make representations to the government to remove these barriers.

Woking Borough Council acknowledges that some specific sustainable energy proposals have the potential to provoke an adverse reaction from some members of the communities they serve. Hence, care has been taken to build an understanding of the concerns that people may have through engagement at very early stages in project development.

The rationale behind community services is to provide a customer-centric outcome-based approach to the services that affect the day-to-day lives of people in the community: that is, to think commercially in securing long-term benefits for the community. For these core services in the community the approach has been to be more open and experimental. The need for improved outcomes and cost-effectiveness has been paramount, and the service areas will be encouraged to take more innovative approaches to securing better outcomes; some will fail and must be taken as learning opportunities not as causes for condemnation. The oversight arrangements by Woking Borough Council's Executive and Climate Change Working Group continue to drive policy innovation and encourage and support the delivery of more customer-centric outcomes.

6.3 Potential for diffusion and scaling up

In progressing a number of sustainability schemes in parallel, Thameswey will be able to become the electricity provider to each of the 1,000 homes joining the scheme, provide each householder with assistance in implementing the power generation from a combination of ground-source heat-pump and photovoltaic electricity while backing up the supply from its fluctuating wind turbines and non-fluctuating CHP electricity generation elsewhere in the Borough.

Thameswey can provide local sustainable energy services to other local authorities, public bodies and the private sector both within and outside Woking. For projects outside of the Borough of Woking the company policy is to establish a subsidiary so that there is a clear division between activity for the direct benefit of Woking and activity that derives an indirect benefit.

In recognition of its work on climate change and sustainable energy, Woking is the only borough that has been invited to join the board of ecoSE, which is leading on the development of good practice for sustainable development in the UK South-East region. Woking is also a member of the South-East Climate Change Partnership, working with the South-East England Regional Assembly (SEERA) on climate-proofing the South-East Plan.

Woking is also a member of the UKCIP Local Authority Working Group, contributing to the development of indicators for climate change and adaptation for local authorities.

6.4 Overall conclusions

There is a need to address climate change corporately and to demonstrate to local authorities the components of success in this: political ownership, technical skills and a sound managerial approach. Woking's achievements in sustainable local energy owe much to the political leadership of the Council in championing the issues of energy efficiency and sustainability, mainstreaming sustainability throughout the Council's business activities and establishing a Climate Change Strategy, maintained through the cross-party ownership of the Council's Climate Change Working Group. The Climate Change Strategy defines specific, measurable local objectives towards challenging national targets for CO_2 emission reduction, with a clear action plan and monitoring system—decoupled from short-term political cycles. The flexible and comprehensive nature of the Strategy with a long-term outlook, covering all of the Borough's energy uses and details on short and medium term actions and progress, in one document, is considered useful (Curran 2007).

Local sustainable energy can be small- or large-scale, and Woking is able to show other local authorities and businesses how to deliver at both scales and how to achieve CO_2 savings. Woking's approach to local sustainable energy systems is to supply customers on private-wire CHP and/or renewable energy networks, as well as implementing energy and environmental services in the public and the private sectors. Thameswey Ltd can indeed provide local sustainable energy services to other local authorities, public bodies and the private sector both within and outside Woking to enable replication.

Resilience to climate change should be designed into new developments by making them less dependent on grid-distributed energy and fossil fuels. The energy consumed by new development should meet the target of reducing greenhouse gas emissions to a level equivalent to 80% of CO_2 emissions compared with the previous use of the land (in 1990). The energy requirements of new development should be met through sustainable energy and renewable energy generation.

Achieving these objectives is promoted through Woking's Climate Neutral Development Guidance, believed to be the first of its kind in the UK. Through ecoSE, which is leading on the development of good practice for sustainable development in the South-East region, membership South East Climate Change Partnership, working with SEERA on climate-proofing the South-East Plan and membership of the UKCIP Local Authority Working Group, Woking is contributing to the diffusion of more sustainable local energy solutions.

A key factor in the Council's success is the combination of technical innovation, such as CHP, absorption cooling, private-wire systems, etc. as well as partnership with the private sector and financial and commercial innovation. For residents, a great benefit is that Woking achieves these objectives in a way that is self-funding, working with commercial and non-commercial partners to meet environmental goals.

Simultaneously, Woking has sought to tackle fuel poverty by reducing heating costs for Council accommodation to within 10% of the state pension and has pursued energy conservation. The aim for such community services is to provide a customer-centric outcome-based approach and to think commercially in securing long-term benefits for the community, also being more open and experimental.

Woking has demonstrated excellent practice in the field of partnership—many of its achievements only being possible via interagency working. The Council also takes care to build an understanding of the concerns provoked by some specific sustainable energy proposals, through engagement at an early stage in project development.

Woking Borough Council's leadership in the UK in the development of local sustainable community energy has been recognised in receiving the Queen's Award for Enterprise: Sustainable Development 2001, the only local authority ever to receive a Queen's Award for Enterprise.

Finally, Woking can share experience and expertise in energy efficiency, with the audience being not only national and local authorities but also private individuals, business and community groups throughout the UK and Europe through ECSC (Unit 327, 30 Great Guildford Street, London, SE1 0HS, UK).

References

IDEA (2005) www.woking.gov.uk/environment/climate/greeninitiatives/beacon (accessed September 2010).

—— (2007) www.localinnovation.idea.gov.uk/idk/core/page.do?pageId=17452596 (accessed September 2010).

—— (2008) www.woking.gov.uk/environment/climate/greeninitiatives/beacon/themeguide (accessed September 2010).

WBC (Woking Borough Council) (2008) 'Climate Change Strategy: A Climate Change Strategy for Woking'; www.woking.gov.uk/environment/climatechangestrategy (accessed October 2009).

7
Energy-saving performance contracting for federally owned public buildings: success factors from the Austrian perspective

Ingrid Kaltenegger
Joanneum Research, Graz, Austria

Angelika Tisch
Inter-university Research Centre, Graz, Austria

In the course of our research project (Tisch *et al.* 2008) we tried to identify and analyse barriers linked to the implementation of product–service systems (PSSs) in public procurement in Austria and strategies to deal with them. Although several studies proved that PSSs are able to offer attractive solutions within the scope of sustainable development, currently, in public administration, PSSs are used only to a small extent.

The first step within the project was to collect and analyse opinions and experiences made by public authorities and companies. In-depth interviews with 20 officers responsible for public procurement and 10 employees of companies that offer PSSs were carried out. The results of these interviews were used to develop two questionnaires, one was sent to more than 3,000 Austrian public procurers, and the other was distributed to 250 companies. A total of 88 responses from public procurers and 35 responses from companies were received. Based on the results of the interviews and the questionnaires, we identified success and hindering factors for the implementation of PSSs in public procurement. Furthermore, we developed strategies to overcome the hindering factors and discussed them with users and providers of selected PSSs in three round-table meetings.

- The federal government, provincial governments or certain public authorities should develop and offer professional assistance for the implementation of PSSs in the field of public administration

- The budgeting should guarantee a benefit from potential financial savings for the users of the PSS

- Joint procurement in public administration should be encouraged

- In the call for tender, procurement officers should choose the criteria of the 'economically most advantageous tender' and consider life-cycle costing

- The public administration should state clearly the persons responsible for the procurement of specific products and services

- Companies should offer intensive support during the implementation of PSSs and intensify the information about the systems they offer

During our research, we discovered an extremely successful example of PSSs where several strategies for the management of barriers were effectively adopted: energy-saving performance contracting (ESPC) for federally owned public buildings in Austria.

This chapter starts with a description of ESPC, a system that becomes more and more attractive. Then, the contracting initiative for federally owned public buildings in Austria is outlined.

Three of the main factors of success are:

- A tender that is designed to select a well-experienced and powerful contractor and a contract that guarantees the participation of users

- A concept that highlights the advantages contracting offers to the user

- An independent third party

The final factor seems to be the most important, as the third party is able to win new users to the contract. Furthermore, it can support the user in the selection of the right contractor and in the implementation of the contracting system

7.1 Energy-saving performance contracting

According to the definition in our project, a PSS incorporates products and services that are designed to fulfil the user's needs in an optimal way. The companies that offer PSS focus on the sale of the desired benefit and not on the sale of a product. In order to achieve this, they change the perspective and create new solutions. PSSs are particularly interesting with regard to their potential environmental effects when compared with those associated solely with the purchase of a product.

ESPC is an example of a PSS. It is an agreement between the owner of the building and the energy service company (contractor) that guarantees a specified saving of energy. The contractor finances, installs and maintains activities to reduce the energy consumption of the building. Principally, the contractor is paid according to the energy that it provides. The payment of the contractor should equal the energy savings that result from the new contract. At the end of the contract, the customer owns the renewed equipment and the benefits of all further savings. Energy performance contracting is an extremely effective way to reduce energy use and costs and to renew facilities and building systems without expending capital funds.

7.1.1 The Austrian contracting initiative

In the past, owing to a tight budget, the federal government in Austria often had to put on hold energy-saving investments in federal buildings. The federal government was not able to provide the budgetary funds in the required amount at short notice.[1]

In 1997 the idea to implement ESPC in federal schools in Vienna was formulated on the basis of a pilot project. After a first analysis of specific energy indices and the condition of the school buildings, two pools with a total of 46 schools were chosen for the pilot project. In 1998 a contract was signed with two contractors that guaranteed a reduction of 24.3% (pool 1) and 21.1% (pool 2) of energy use, respectively.

In 2003, after positive results from the pilot project, the federal government decided to expand the contracting initiative. Currently, numerous Austrian schools, ministries, barracks and universities have a contractor that will be informed in case of unexpected events, for example when the school bell does not ring.

For numerous buildings, operating maintenance has already been eliminated for cost reasons. In recent years, the majority of public buildings (approximately 75% of the total number of such buildings) have been transferred into commercial ownership of the Bundesimmobiliengesellschaft (BIG).[2]

The successful pilot project was followed by a decision of the Council of Ministers in 2001 to renovate approximately 300 properties by ESPC. Appropriate energy-saving partners had to be found for these buildings by the end of 2004; these partners had to identify measures to save energy, implement, pre-finance and manage the buildings for 10 years. The 300 properties belong to the following agencies of the federal government:

- Federal Ministry for Education, Science and Culture
- Federal Ministry for Finance
- Federal Ministry for Justice
- Federal Ministry for the Interior
- Federal Ministry for Economy and Labour
- Federal Chancellor's Office
- Parliament

1 See www.energhyagency.at/fileadmin/aea/pdf/publikationen/energy/energy-01-2002.pdf (accessed 8 November 2010).
2 *Ibid.*

The project is being handled by a cooperative association of the Federal Ministry for Economy and Labour (BMWA), the Federal Ministry for Land and Forest Management, Environmental and Water Management (BMLFUW) and the BIG. The cooperative association will be advised in specialist and organisational measures by the Austrian Energy Agency (EVA).

The building investments, which for numerous reasons would not have been performed without ESPC, offer a series of positive economic and ecological effects, as stated in the ministries' paper on ESPC.[3]

The federal government—as building owner and user—saves approximately €1.4 million in energy costs per year over the 10-year contract because it experiences savings immediately. After the contract has terminated, the federal government enjoys the benefits of all the energy savings (approximately €6.9 million per year). Because ESPC involves a service that also includes technical operations, inspection and some maintenance tasks, significant savings also arise in this area. An estimate, made by the project group of the contracting offensive, shows annual cost savings in the magnitude of €2 million to €3 million in the operation, inspection and maintenance of the technical energy systems alone. After the expiration of the contract, it can be expected that the costs for these items will rise again as the contractor is no longer responsible for the systems and separate maintenance contracts will have to be started once again.

As the ministries who initiated this contracting-offensive have calculated, the implementation of energy-saving measures in the buildings makes a significant contribution to fulfilling the climate protection obligation of the Republic of Austria (a 13% reduction in carbon dioxide [CO_2] emissions compared with 1990). Through the contracting offensive, 10% of the CO_2 emissions will be saved among all BIG buildings used by the federal government. Through ongoing general renovations of additional federal buildings, another 7% of the CO_2 emissions can be reduced during forthcoming years. As such, the federal government will fulfil its role as a model for private building managers.

The partners contractually agree on the maintaining of a pre-decided comfort level (inside room temperature, lighting strength and air exchange rate). Use of the building will not be limited by ESPC, but if any change in use exceeds a specific scope, the agreement provides that the initial conditions must be corresponding recalculated.

7.1.2 Success factors

The example of ESPC for federally owned public buildings offers several strategies to overcome hindering factors for PSSs, both for those public authorities that are already using PSSs as well as for those that do not. In the following, three main strategies or success factors are presented that were both mentioned during our in-depth interviews and the questionnaires as well as during the round-table discussions. All three strategies are pursued in the Austrian contracting initiative.

3 *Ibid.*

7.1.2.1 The tender ensures a premium contractor and the participation of the user

Evaluation of the questionnaires indicates that the long-term bond to the contractor is the most significant factor hindering implementation of PSSs for those public authorities who already use these systems. At the same time, as one of the experts pointed out, the custodians of public buildings often require a contact person to whom they can relate, because otherwise most of them tend to be overstrained. For example, the school custodian could be overburdened with the variety of questions concerning the heating system, the thermal insulation of the building or the behaviour of the occupants concerning energy efficiency. Therefore, it seems that in the field of energy contracting the core problem is not the long-term bond to the contractor but the bond to a contractor who is not competent enough and the absence of possibilities for the occupants of the building to take part in the decisions of the contractor. As one of our experts put it, much is destroyed by choosing the 'wrong' contractor, not only in the public authority itself but also in those authorities with which that public authority is in contact.

In Austria, the tender and the contract are designed according to certain principles, in order to ensure a premium contractor and the possibility of user participation. So-called negotiated procedures are chosen to award the contract. In the first step of this procedure, appropriate bidders are selected. In a second step, those bidders are invited to tender. It is important to mention that the offers are not compared on the basis of the lowest price alone. The contract is awarded to the economically most advantageous tender that was able to meet other award criteria such as practical experience or ability to motivate the occupants of the building to save energy (for example, by displaying a board that shows the current energy use in the building, the employment of 'energy sheriffs' who make sure that, for example, the lights are switched off when rooms are not in use or setting up regular meetings with operating staff such as the custodians). The best offer is chosen by a jury that includes occupants of the building.

The contract, which runs over a period of 10 years, is based on a model contract developed by the Austrian Energy Agency, a science-based union, widely appreciated in Austria. It makes high demands on the contractor and establishes, for example, comfort conditions such as room temperature. In the contract, the possibilities for participation of occupants are defined, too. For example, the ways in which the occupants can influence decisions of the contractor where to invest in the building are fixed in the contract. This helps to overcome barriers such as fear of low temperatures or fear of 'being at the mercy' of the contractor.

7.1.2.2 The concept highlights the advantages for the user

The questionnaires sent to officers responsible for public procurement showed that nearly 66% of those who did not use PSSs were not able to recognise their benefits. The Austrian contracting initiative deals with this problem by pointing out the advantages to the users of the buildings. It is not only that the initiative informs the potential users about the benefits but that the whole initiative is designed in a way that the advantages offered are diverse and can easily be recognised:

- Users profit from a 20% reduction in energy costs

- Users profit from the upgrading of their building and of its technical systems

- The users obtain partners that guarantee a certain outcome (for example, a minimum temperature for the swimming pool); in case of serious technical difficulties, the contractors are obliged to solve the problem within a couple of hours

- The contractor implements permanent measures that aim at encouraging and motivating the occupants to save energy; by changing the daily habits of the occupants of the building, further energy savings in their private homes are possible

7.1.2.3 There is an independent third party that has the trust of the user

The evaluation of the questionnaires of those public authorities that did not use PSSs shows two main hindering factors: it is feared the tendering procedure is too long and complex and that there will be a lack of external assistance. Both hindering factors can be overcome with the help of an independent third party that the users trust. In the field of ESPC for federally owned public buildings the existence of such a third party seems to be the most significant factor for success.

In this particular case, it is the Ministry of Deployment and Economy, with its energy specialists assuming the role of the independent party. The role of these specialists, launched by the Ministry at the end of the 1970s, is to support government departments in saving energy. They recorded the heating systems and other energy technologies, identified energy indices and suggested improvements. Therefore, they were in contact with the relevant persons in most of the federally owned buildings and normally enjoyed their confidence and also knew quite well the buildings and their problems regarding energy savings.

In the course of the ESPC they are assigned various functions:

- To inform and motivate: in the run up to the tender procedure, the energy specialists carry out talks and workshops with those governmental departments chosen potentially to participate in the contracting initiative. They reveal the benefits of the ESPC and inform the potential participants of the steps to be taken and possible results. They try to diminish potential participants' fear of this unknown system (as this is a further hindering factor mentioned by an expert)

- To carry out the tender procedure and give support: after the governmental departments decide to take part in the contracting initiative, the energy specialists carry out the call for tenders. Furthermore, they support the users during the operating time of the contract by controlling the account of the contractor or by organising annual meetings between contractor and users

- To act as arbitrator: in case of conflicts between users and contractor, they act as arbitrator

With the decision to assign energy specialists with the promotion and management (monitoring, controlling, etc.) of the ESPC, the Ministry of Employment and Economy not only chose trusted and well-informed managers but also made sure that those people would not lose their employment in the process of the implementation of the PSS.

In this section we have discussed the three main success factors that are of special importance for the ESPC. There are many more success factors in the implementation of PSSs in general, a few of which will be mentioned below in Section 7.1.3.

7.1.3 Lessons learned for the implementation of other product–service systems in public authorities

It seems that especially for those PSSs that are more complex and innovative than, for example, copier service systems, public authorities need a third party that informs, motivates and supports (potential) users and acts as intermediary between users and providers. Not only during the introduction of the PSS but also during its execution. It is important that this third party is trusted not only by the users but also by the providers. Therefore an independent and competent third party should be chosen.

Further findings in Austria strengthen the importance of a third party in the course of the implementation of PSSs in public authorities. Currently, there is a slight push towards PSSs in public authorities in Austria that originates from the central procurement agency. This agency has supported more recent PSSs that offer cost benefits or legal certainty (such as printer service systems or car-pools). To some extent, the central procurement agency acts as a third party; it reveals the benefits, informs about the PSS and could even assume the role of the arbitrator in the case of differences. Therefore, those who want to implement PSSs in public authorities should look for an institution that is able to act as a third party.

This conclusion could be of importance for PSSs in private procurement, too. Recent research regarding the implementation of new PSSs reveals problems (Rabelt *et al.* 2007). Even where the concept was convincing, several PSSs were unable to 'gain ground'. The importance of a third party as an intermediary is illustrated by research projects (for example, see Ax and Becker 2007; Faltz *et al.* 2007) where the research institute (unwillingly) took on the role of third party to facilitate the process. There are cases where the PSSs lasted only as long as the life-span of the project. It seems that in discussion so far about PSSs the importance of a third party that permanently supports the system is underestimated.

In the case of the ESPC a tender procedure was chosen, which aims at guaranteeing a highly qualified provider. The results of our interviews and round-tables show that this is already taken into account by procurement officers who procure other PSSs such as the rental of textiles for medical use. In some cases it seems to be difficult to define the quality of the providers' performance in the tender in a way that ensures that the company that offers the best quality is chosen.

Tenders that do not address PSSs exclusively should consider a couple of points to allow PSSs to be taken into account:

- Use performance-based or functional specifications
- Take account of alternatives that can be submitted by the participants in the tender
- Use a life-cycle costing approach
- Award the contract to the economically most advantageous tender and not to the tender with the lowest price

The current initiatives of the EU and its member states (for example, see BMWA 2007) aim to support innovations through public procurement. They make suggestions for the legal and procedural framework of procurement procedures. Once applied in the day-to-day business of procurers, it should be easier for PSSs to be 'captured' by public authorities.

The separate cost accounting is the basic principle that allows procurement officers to identify the potential cost benefits of PSSs. In the case of ESPC the energy specialists make sure that the life-cycle costs of the energy supply in federally owned public buildings are evident. The providers of other PSSs are less fortunate. For example, many smaller public authorities are not aware of the precise costs of their street lighting. Therefore they are not able to evaluate whether a contracting solution for street lighting offers any cost benefits. It seems that with the current initiative of the EU mentioned above, the separate acquisition of costs and a life-cycle costing approach are going to gain ground in importance in public authorities.

7.2 Overall conclusions

The example of the Austrian contracting initiative for public buildings makes clear that in some cases public authorities are able to lead the way towards sustainable development.

In the case described in this chapter the public authority presented a solution that not only reduces energy consumption and costs but also offers an appealing concept that is able to overcome difficulties and offer diverse advantages. After the life-span of the contract, which is 10 years, the public authorities will have gained wide experience in the field of efficient energy supply to buildings. In addition there is hope that the occupants of the buildings will behave more energy efficiently outside the workplace.

Six main recommendations in general can be summarised for a successful implementation of PSSs in public procurement:

- To support the implementation of PSSs in public procurement, qualified public bodies should develop and offer professional support. The model of energy performance contracting can serve as a model that also shows the innovative ability of public administration
- Public administration should have access to more information about PSSs, provided ideally by state and federal government

- Officers in public procurement should, whenever possible, look for a premium contractor and consider the life-cycle costs

- Public procurement should make responsibilities for procurement more transparent

- Providers should broaden their offer to support the public administration in the implementation of PSSs and provide more information about it

- If possible, budgeting should be designed in a way that the users profit from their efforts. The implementation of further PSSs in public authorities would be easier, if the users or occupants could profit directly from the savings. In the contracting initiative the users of the building currently profit only in an indirect way

The need for professional support for the implementation of PSSs plays an important role for public administration, but it is quite probable that it is also important for private users and business. For further research it would therefore be wise to look for answers to the following questions:

- What are the special requirements for different building ownership types (public administration, private households, enterprises) that have to be satisfied by a professional provider and what organisations could fulfil this role?

- What measures have to be taken within the scope of professional support?

- How can all relevant actors be involved in an optimum way in the implementation of product–service systems such as energy-saving performance contracting?

References

Ax, C., and F. Becker (2007) 'ReUse: Regionale Netzwerke für eine nachhaltige Nutzung von Computern', in V. Rabelt, A. Heimerl, K.H. Simon and I. Weller (eds.), *nachhaltiger_nutzen. Möglichkeiten und Grenzen neuer Nutzungsstrategien* (Munich: oekom verlag): 36-50.

BMWA (Bundesministerium für Wirtschaft, Familie und Jugend) (2007) *procure_inno: Praxisorientierter Leitfaden für ein innovationsförderndes öffentliches Beschaffungs- und Vergabewesen* (Vienna: BMWA; www.bmwa.gv.at [accessed October 2009]).

Faltz, L., W. Baumann, D. Drenk and K. Hesse (2007) 'ecomöbel: Durch Kooperation zur Wiederverwendung von Gebrauchtmöbeln; Design und Ökologie im Einklang', in V. Rabelt, A. Heimerl, K.H. Simon and I. Weller (eds.), *nachhaltiger_nutzen. Möglichkeiten und Grenzen neuer Nutzungsstrategien* (Munich: oekom verlag): 16-35.

Rabelt, V., A. Heimerl, K.H. Simon and I. Weller (eds.) (2007) *nachhaltiger_nutzen. Möglichkeiten und Grenzen neuer Nutzungsstrategien* (Munich: oekom verlag).

Tisch, A., I. Kaltenegger and A. Windsperger (2008) 'PSS-ÖB: Strategies to Deal with Barriers to the Implementation of Ecoefficient Product–Service Systems in Public Procurement', project report; www.fabrikderzukunft.at/publikationen/view.html/id662 (in German; accessed 8 November 2010).

8

Building Investment Decision Support (BIDS™) for green building technologies

Vivian Loftness, Volker Hartkopf, Azizan Aziz, Megan Snyder,
Joonho Choi and Xiaodi Yang
Carnegie Mellon University, USA

8.1 Moving beyond broad definitions of sustainability to justify high-performance materials and assemblies

Investment in high-performance, sustainable building solutions and technologies is limited by first-cost decision-making. In our collective enthusiasm to define and promote sustainability, we may be making two fundamental errors: first, by using broad generalised definitions of sustainability, and second, by making arguments that 'green' design need not cost more.

Rather than broad sustainability objectives, investors and clients will need to understand the quality differences of sustainable design alternatives—component by component—if they are to move beyond least-first-cost decision-making. While 'mobility' can be purchased in cars as inexpensive as $10,000,[1] every investor knows component by component the quality differences in $10,000 and $30,000 cars, and predominantly invests in the higher-cost, higher-quality solution. While life-cycle decision-making is integral to a five-year car purchase, buildings are still built on a least-first-cost basis, despite the 30–50 year life expectation. The genius of LEED™ certification from the US Green Building Council[2] is that it defines sustainability in 69 distinct goals, giving the

---

1 All dollar amounts are in US dollars.
2 www.usgbc.org

client the opportunity to qualify a greater investment of expertise and capital in their buildings.

While promoting either broad or detailed sustainability goals, many sustainable building designers will simultaneously argue that 'green' design should not cost more. A number of US studies have demonstrated that modest cost increases can achieve silver-level and gold-level LEED™ certification, achieving higher levels of sustainability with short-term paybacks (Davis Langdon 2007; Kats 2009). Although invaluable arguments for introducing sustainability, these modest cost increases are locking architects and engineers out of true quality improvements in a wide range of building materials, components and systems that are critical to ensuring indoor air quality, thermal control, lighting control, network access, privacy and interaction, ergonomics and access to the natural environment. The 10-year effort of the Advanced Building Systems Integration Consortium (ABSIC), a consortium of industries and federal agencies,[3] and the Centre for Building Performance and Diagnostics (CBPD)[4] to define high-performance buildings continues to reveal innovative design options that will greatly enhance the quality of the individual workplace. Box 8.1 outlines the CBPD guidelines for high-performance buildings. The BIDS™ (Building Investment Decision System) life-cycle decision-support tool, to be described in this chapter, continues to build on the return-on-investment (ROI) calculations for these guidelines, based on actual cases.

Box 8.1 Centre for Building Performance and Diagnostics (CBPD) and Advanced Building Systems Integration Consortium (ABSIC) design guidelines for high-performance buildings

Guidelines for high-performance enclosure systems

1. Maximise individual access to the natural environment

2. Maximise daylighting for task and ambient lighting

3. Maximise natural ventilation with mixed-mode conditioning

4. Minimise enclosure heat loss and heat gain

5. Design solar heat and glare control

6. Engineer load balancing and mean radiant temperature control

7. Engineer passive and active solar heating, cooling and power

8. Maximise enclosure integrity and material sustainability

9. Innovate through systems integration

---

3 ABSIC members: Armstrong World Industries Inc., BP Solar, Carnegie Mellon University, US Department of Energy, US Department of Defence, Electricité de France, US Environmental Protection Agency, US General Services Administration, Public Works and Government Services Canada, Siemens Energy and Automation Inc., Steelcase Inc., Teknion Inc., Thyssen Krupp AG, Tyco Electronics, United Technologies/Carrier and the National Science Foundation.

4 Centre for Building Performance and Diagnostics, School of Architecture, Carnegie Mellon University, 5000 Forbes Avene, Pittsburgh, PA 15213-2890, USA.

Guidelines for high-performance heating, ventilation and air conditioning (HVAC)

1. Use separate ventilation systems from those used for thermal conditioning
2. Design for natural ventilation with mixed-mode conditioning
3. Provide task conditioning and individual control
4. Design for continuous change with modular, integrated HVAC and controls
5. Design architecture 'unplugged' for maximum efficiency and passive
6. Engineer load balancing
7. Engineer energy-effective and material-effective HVAC systems with 'energy cascades'
8. Create distributed, communicating, modifiable automation systems with individual control
9. Seek innovative HVAC system integration for thermal and air quality, resource conservation and environmental health

Guidelines for high-performance lighting

1. Provide daylighting as a dominant light source
2. Separate task lighting from ambient lighting or design relocatable task–ambient systems
3. Introduce indirect–direct lighting to support spatial dynamics without shadowing
4. Maximise lighting quality with high-performance luminaires
5. Provide for reconfigurability with plug-and-play fixtures
6. Design for continuous change in lighting zone size and advanced controls
7. Pursue innovative systems integration

Guidelines for high-performance connectivity

1. Engineer independent plug-and-play networks—data or voice, power, security and environmental services—with central communication
2. Design distributed cores for accessible, modifiable vertical distribution
3. Design distributed satellite closets with plug-and-play interfaces
4. Resolve integrated, reconfigurable plenum systems (ceiling or floor)
5. Ensure user-accessible, modifiable grid and nodes of services for connectivity
6. Create wiring harnesses for data or voice, power, security and environment
7. Select terminal units that provide all services—data, power, voice, security, environment—in reconfigurable boxes for 'just-in-time' modifications
8. Create robust monitoring and individualised controls

Guidelines for high-performance interior systems

1. Design neighbourhood clarity and shared spaces with flexibility
2. Design layers of ownership, with multiple work environments
3. Ensure ergonomics and functional support for shared work processes
4. Ensure ergonomics and functional support for individual work processes
5. Design 'layers of closure', privacy and acoustic control
6. Design 'layers of mobility' for workstations and workgroups
7. Provide levels of personalisation
8. Ensure environmental infrastructure to support changing densities and closure
9. Ensure technical infrastructure to support changing densities and closure
10. Select interior systems and components for material and energy conservation
11. Select healthy, maintainable interior components
12. Design for access to the natural environment

Design process changes for high-performance buildings

1. Involve the full design team from the outset to ensure integrated design
2. Develop prototyped, modular and integrated delivery of air quality and thermal conditioning systems, as well as lighting/daylighting and connectivity systems
3. Shift from design–build to manufacture–install for life-cycle value
4. Shift to just-in-time purchasing of major building infrastructure for quality with cost control
5. Establish flexible infrastructures for dynamic organisations

---

A high-quality light fixture—one with the most energy-effective T-5 lamp, continuous dimming and daylight responsive ballast, high-performance reflector and lens and potentially even separate ambient uplighting and task downlighting—may indeed cost three times the least-cost 2 × 4 offer so prevalent in our buildings. It is imperative that life-cycle data sets and tools be developed to establish the cost benefits of high-performance building technologies—component by component.

8.2 To justify high-performance building components and systems, understand the cost of ownership

In order to promote investment in sustainable, high-quality buildings, it is critical to prove to the client that the real cost of doing business is realised over time, not in first construction costs. Careful book-keeping will reveal that 'cheap' buildings and infrastructures, and 'cheap' building delivery processes, result in major costs over time—energy costs, waste and renewal costs, as well as productivity and health costs.

Moving well beyond the 'mantra' of individual productivity, the CBPD team has been researching the range of workplace-related expenses that are carried annually by organisations—from energy and facility management costs to spatial churn to staff turnover, health and litigation costs. Many professionals know about the comparative costs of productivity at US$2,000 per square meter per year, rent at US$200 per square meter per year, and energy at US$20 per square meter per year (2008 dollars). Yet productivity in the white-collar workplace is hard to define and difficult to measure. As a result, arguments for high-performance, sustainable buildings may be more convincingly made with other annual expenses carried by the organisation. The Carnegie Mellon BIDS™ research team has identified a list of ten cost benefit areas where annual organisational investment is significant and could be reduced through a commitment to higher quality buildings (Box 8.2).

Box 8.2 The benefits of high-performance buildings, according to the Centre for Building Performance and Diagnostics BIDS™ (Building Investment Decision System) research team

First-cost and mortgage savings through quality packages

- Integrated system savings over individual components
- Quality and modularity benefits with just-in-time purchasing over redundancy

Facilities management cost savings

- Savings in maintenance and repair
- Savings in energy, water and other utilities
- Reduced discomfort complaints and associated costs
- Reduced costs for failure
- Benefits from employee retention and training

Individual productivity cost savings (skill-based, rule-based and knowledge-based jobs)

- Increase in speed and accuracy
- Increased effectiveness, creativity and motivation
- Decrease in absenteeism

Organisational productivity cost savings

- Increased profit
- Reduced time to market
- Increased customer attraction and retention
- Recognition and publicity
- Continuous work flow

- Real estate value (or income?)
- Increased effectiveness
- Increased team and multidisciplinary creativity

Attraction and retention or turnover cost savings

- Reduced time and cost to attract employees
- Increased quality of employees attracted
- Reduced training costs
- Increased retention rates

Tax, building code compliance, insurance and litigation cost savings

- Increased utility and tax incentives
- Decreased tax depreciation
- Increased code compliance
- Decreased insurance and litigation costs

Health cost savings

- Decreased compensation to workers
- Decreased medical insurance costs
- Decreased health litigation costs
- Better environmental evaluation and remediation
- Reduction in lost work time

Spatial renewability cost savings

- Organisational churn
- Reduced labour and material costs for reconfiguring workstations and workgroups
- Reduced HVAC, lighting and networking system modification costs
- Reduced occupant down-time

Technological renewability cost savings: technological churn

- Better networking: data, power and voice change
- Improved hardware and software change
- Reduced training and mentoring costs
- Reduced organisational and workspace costs
- Reduced environmental and conditioning response costs

Salvage and waste cost savings

- Reduced cost for organisational, technological, environmental modifications
- Reduced activity-related waste
- Less ageing, wear and obsolescence
- Increased salvage value

Figure 8.1 illustrates the strength of the data that the Centre for Building Performance at Carnegie Mellon University (CMU) has been able to assemble on the cost of doing business that can be critically linked to the quality and sustainability of building materials, components and systems.

8.3 BIDS™: linking quality building components to life-cycle gains

Initiated in 1999, the goals of the BIDS™ project are to develop a cost–benefit analysis framework for various advanced and innovative building systems and to incorporate these within a web-based decision-support tool. There have been four specific objectives set to achieve these goals:

- The development of economic language and logic whereby intelligent workplace design can be thought of by the business investor as analogous to other emerging, strategically central investments that have different operating life-cycles (economic sustainability), competitive implications (workforce impacts) and payback periods (capital market valuation criteria)

- The development of a cost–benefit analysis framework for evaluating various advanced and innovative building system options in relation to a range of cost–benefit or productivity studies, to be incorporated within a web-based decision tool (Figs. 8.2a and 8.2b)

- The determination of cost centres where the benefits of high-performance approaches will be significant, and the expansion of a database relating quality indoor environments to major capital cost and benefit areas, including productivity, health and operations costs

- The identification of laboratory and field case studies demonstrating the relationship of high-performance components, flexible infrastructure and systems integration to the range of cost benefit or productivity indices

Extensive review of the relevant literature to identify valuable case studies as well as related cost–benefit baseline data is a major part of this research project. The CMU BIDS™ team has been avidly pursuing data sets from around the world that link improved building environmental quality to life-cycle cost benefits. With over ten years of research attention, the BIDS™ tool now has over 350 case studies linking high-performance building components and systems to life-cycle value.

With the expansion of the case study database, the BIDS™ tool is beginning to have an adequate number of cases to derive cross-sectional findings in relation to providing air, thermal control, lighting control, network access and access to the natural environment for the individual workplace. With support from the US Department of Energy,

FIGURE 8.1 The true cost of doing business

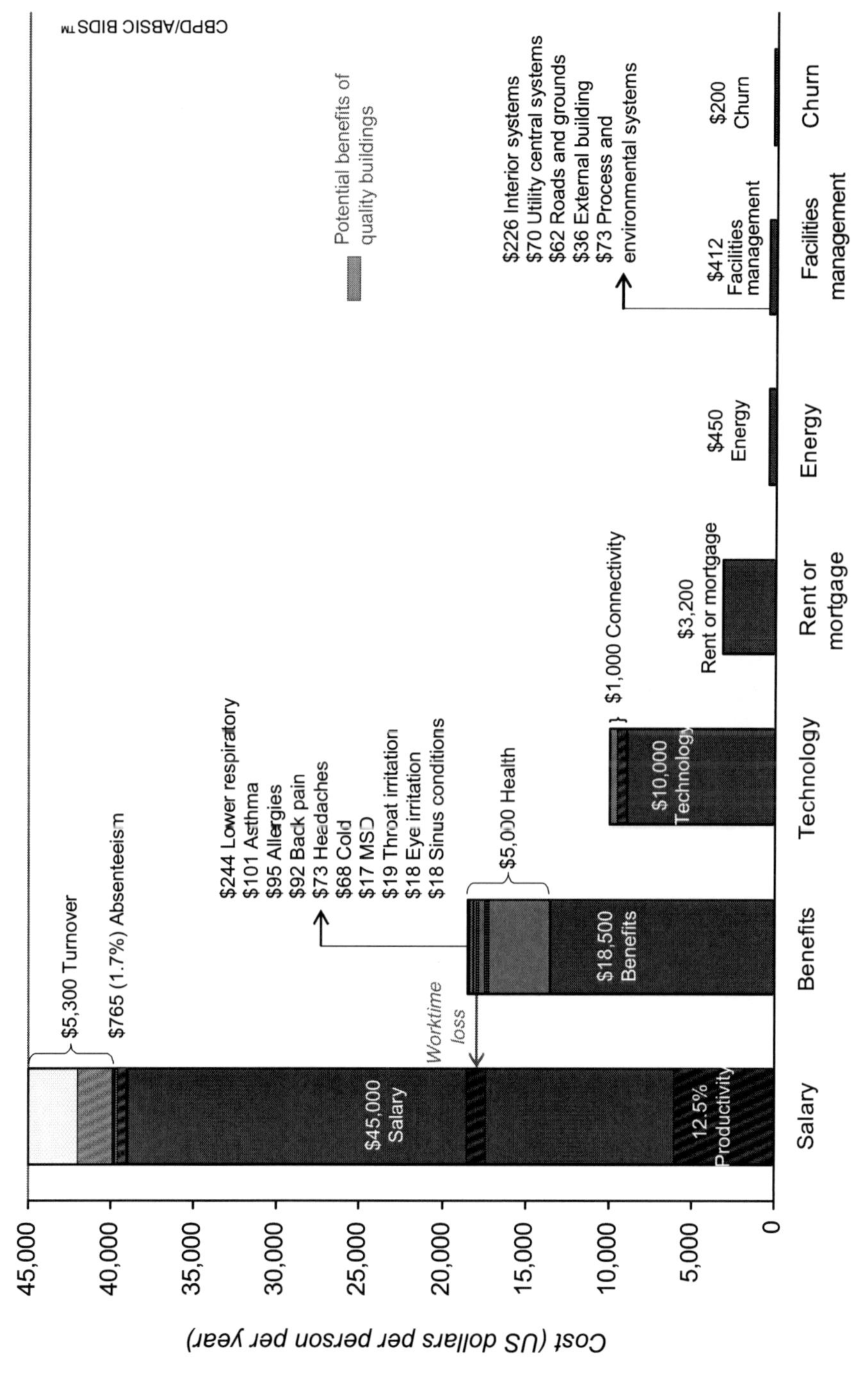

Cost (US dollars per person per year)

CBPD/ABSIC BIDS™

Potential benefits of quality buildings

$5,300 Turnover
$765 (1.7%) Absenteeism

$244 Lower respiratory
$101 Asthma
$95 Allergies
$92 Back pain
$73 Headaches
$68 Cold
$17 MSD
$19 Throat irritation
$18 Eye irritation
$18 Sinus conditions

$5,000 Health

Worktime loss

$45,000 Salary

12.5% Productivity

$18,500 Benefits

$10,000 Technology

$1,000 Connectivity

$3,200 Rent or mortgage

$450 Energy

$412 Facilities management

$226 Interior systems
$70 Utility central systems
$62 Roads and grounds
$36 External building
$73 Process and environmental systems

$200 Churn

Salary   Benefits   Technology   Rent or mortgage   Energy   Facilities management   Churn

45,000
40,000
35,000
30,000
25,000
20,000
15,000
10,000
5,000
0

FIGURE 8.2   (a) The three dimensions of the intelligent workplace BIDS™–EVA matrix and (b) The BIDS™ tool interface

(a)

Design options		Cost–benefit factors		Scenarios
Air Temperature control Lighting control Network access Privacy and interaction Ergonomics Access to environment Whole building	X	First cost O&M, energy Organisational churn Technological churn Individual productivity Organisational productivity Health Attraction and retention of staff Taxes, litigation, insurance Salvage and waste	X	Baseline Globalisation Collaboration Technological dynamics Organizational dynamics Gold-collar orientation Environmental agendas Merger and divestment costs Federal government

(b)

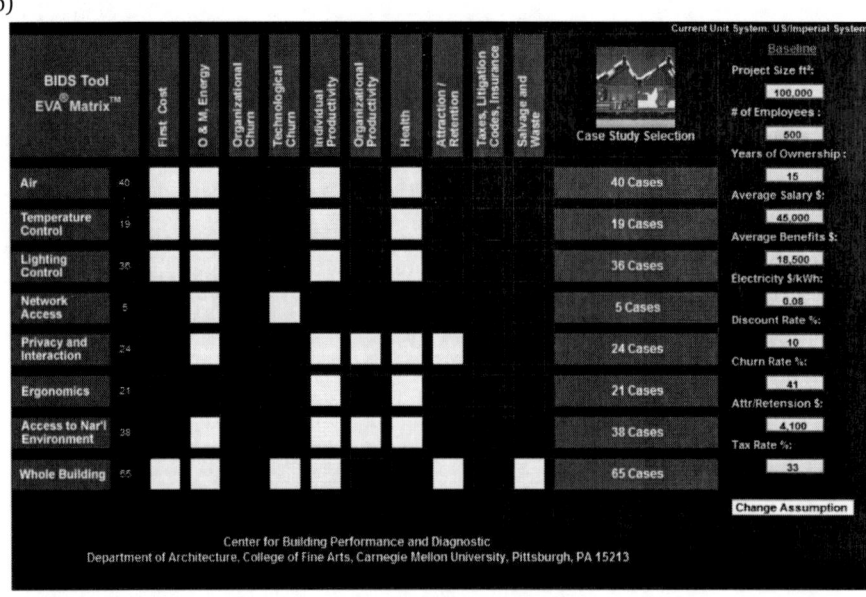

O&M = Operation and maintenance

Source: Centre for Building Performance and Diagnostics, Department of Architecture, College of Fine Arts, Carnegie Mellon University, Pittsburgh, PA 15213, USA

these cross-sectional findings enable us to argue convincingly for critically important improvements in the quality of our buildings.[5]

8.3.1 Access to the natural environment: daylight

Given the dominant number of existing buildings—schools, hospitals, offices and manufacturing facilities—originally designed for effective daylighting and natural ventilation, the erosion of natural conditioning has a serious energy cost. Effective daylighting can yield 10–60% reductions in annual lighting energy consumption, with average energy savings for introducing daylight dimming technologies in existing building at over 30% (Loftness 2005). Emerging mixed-mode HVAC systems that interactively support natural ventilation or air conditioning are demonstrating 40–75% reductions in annual HVAC energy consumption for cooling. Moreover, design for access to the natural environment, including daylighting and natural ventilation strategies, has shown measurable gains for productivity and health in the workplace (Loftness 2002). All high-performance buildings should meet European and Scandinavian standards that ensure that every worker is within seven meters of a window wall, for views, light and air. The effective use of natural conditioning with well designed windows, window controls and mechanical and lighting system interfaces promises to yield major energy efficiency gains, reduced risks in power outages as well as provide measurable health and quality-of-life gains.

A cross-section of BIDS™ case studies indicates that the use of daylight without glare combined with daylight-responsive lighting controls results in 44% average annual whole-building energy savings, 52% average lighting energy savings and an average individual productivity gain of 5.5% (Fig. 8.3). These 11 case studies have shown that innovative daylighting systems can pay for themselves in less than one year as a result of energy and productivity benefits. The BIDS™ tool demonstrates that daylighting yields annual energy cost savings of $112 per employee and annual productivity gains of $2,475 per employee, to give total savings of up to $2,587 per employee annually (Fig. 8.4). With a one-time first-cost premium of $600 per employee (approximately $0.30 per square metre in new construction), the ROI for an investment in daylighting is over 185%.

8.3.2 Access to the natural environment: natural ventilation

Case studies have shown that natural ventilation and mixed-mode HVAC systems that support both air conditioning and natural ventilation provide an average of 59% HVAC energy savings, 9% productivity gains and 1% health cost savings, with less than one-year paybacks. The BIDS™ assembled case studies demonstrate that natural ventilation and mixed-mode HVAC systems yield annual energy cost savings of $110 per employee, health cost savings of $60 per employee and annual productivity gains of $3,900 per employee, for a total saving of $4,070 per employee annually (Fig. 8.5). With an esti-

---

5 See cbpd.arc.cmu.edu/ebids (accessed October 2009).

FIGURE 8.3 (a) Energy and (b) productivity benefits of daylighting

(a)

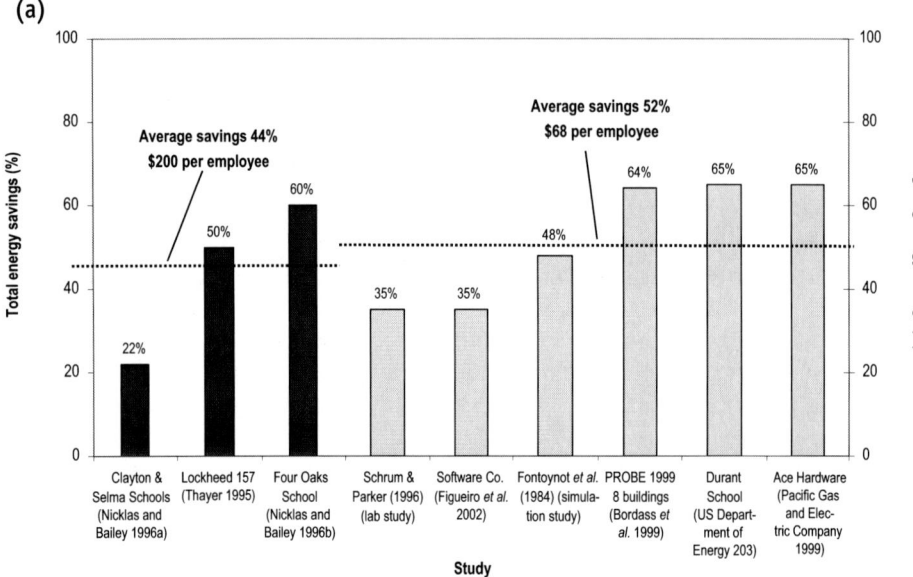

(b)

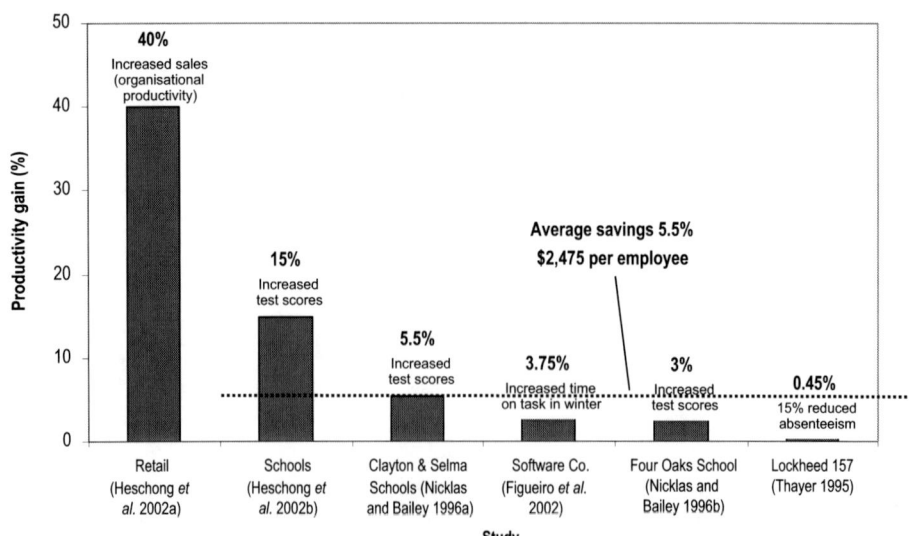

FIGURE 8.4 Costs and benefits of high-performance daylighting

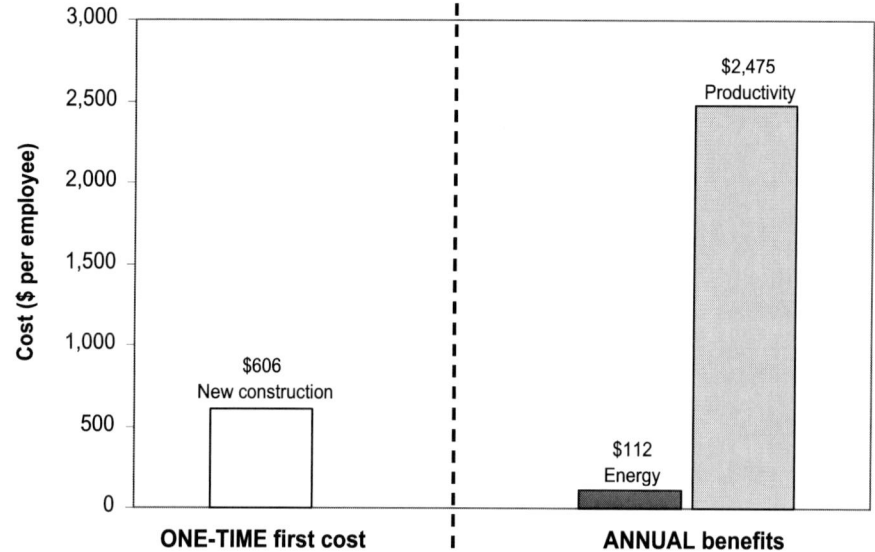

FIGURE 8.5 Productivity gains from mixed-mode conditioning and natural ventilation

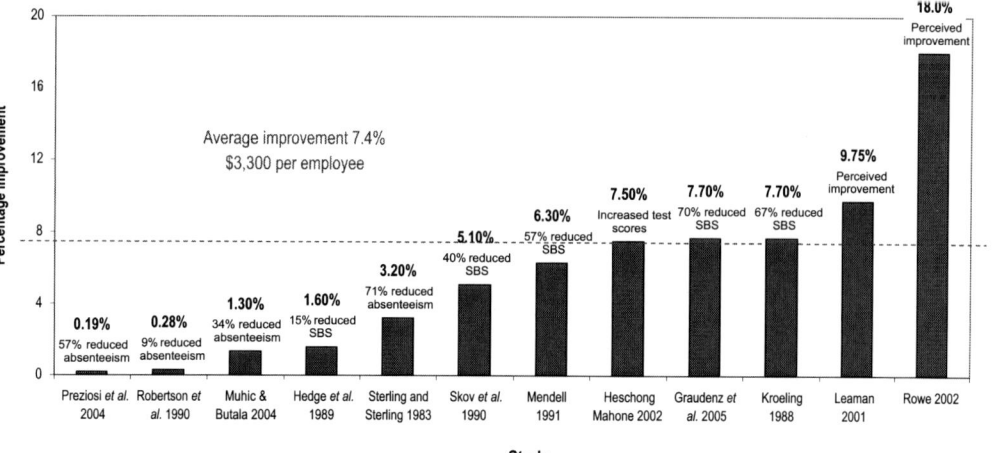

mated first-cost premium of $1,000 per employee in new construction to create narrower buildings that provide every employee with seated views, the average ROI for an investment in natural ventilation or mixed-mode conditioning is 407% for new construction and 120% for retrofits (Fig. 8.6).

FIGURE 8.6 Costs and benefits of mixed-mode conditioning and natural ventilation

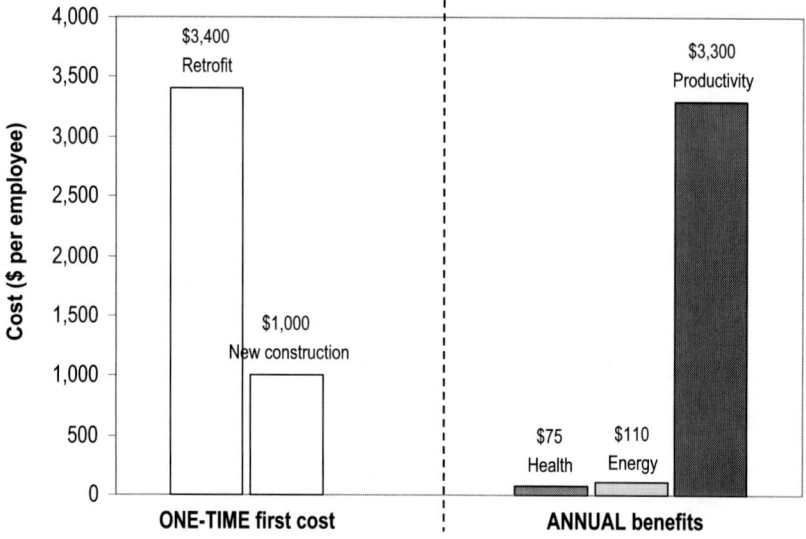

8.3.3 Innovative technologies: high-performance lighting

Value engineering exercises will often trade the quality and efficiency of the equipment and appliances specified for first-cost savings. Even short-term energy savings do not seem to be enough to drive decision-makers towards quality. Either performance standards or links to productivity, health and other life-cycle variables will be critical for promoting investments in high-performance equipment.

In the USA the introduction of national standards for equipment and appliance efficiency has had a major impact on national energy use, reducing overall energy consumption for heating, cooling and refrigeration by 25%, 40% and 75%, respectively (Rosenfeld *et al.* 2004). The direct relationship of appliance electricity demand and CO_2 production illustrates the value of these energy savings in addressing climate change and reducing pollution from power plants. The impact of research and development coupled with standards has enabled refrigerator size and amenities to increase while overall energy use is reduced (Rosenfeld 2003). Four newer appliance standards (washing machines, fluorescent light ballasts, water heaters and central air conditioners) are projected to save US consumers $10 billion in unnecessary energy costs, improve functionality and reduce cumulative emissions by as much as 80 Tg CO_2 equivalent through 2010 (USDS 2002). Given the natural replacement cycle of building appliances and equipment, 190 billion kWh of power demand can be eliminated by 2010 and another 130 billion kWh can be eliminated by 2050 by just four building technologies—ballasts, lamps, windows, and refrigerators and freezers. There are few engineering obstacles and significant export growth potential in expanding appliance and equipment energy efficiency standards to cover the full range of existing and new equipment being introduced in residential and commercial buildings. Barring this commitment from the federal government or states, however, practitioners will need to use every life-cycle value in their promotion of high -performance technologies.

Some 25 studies have helped to quantify the assertion that high-performance lighting systems can pay for themselves in less than a year as a result of the energy, productivity and health benefits. These studies demonstrate an 18% average annual whole-building energy savings, 60% average annual lighting energy savings and an average 3% productivity gain. Use of high-performance lighting fixtures that support daylighting additionally yields annual energy savings of $82 per employee, annual productivity gains of $1,600 per employee, and annual health cost savings of $20 per employee, giving total savings of up to $1,702 per employee annually (Fig. 8.7). With a median first cost of $720 per employee for lighting retrofits, and a median first-cost increase of $200 per employee for high-performance lighting systems in new construction, an investment in high-performance electric lighting results in a ROI of 236$ for retrofits and 851% for new construction (Fig. 8.8).

FIGURE 8.7 Costs and benefits of high-performance electric lighting systems

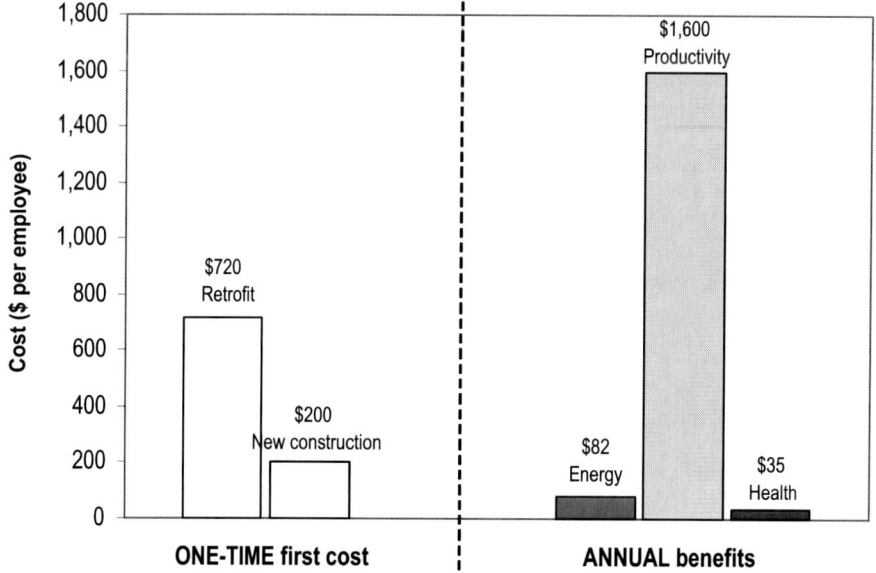

FIGURE 8.8 Percentage productivity gains from high-performance lighting systems

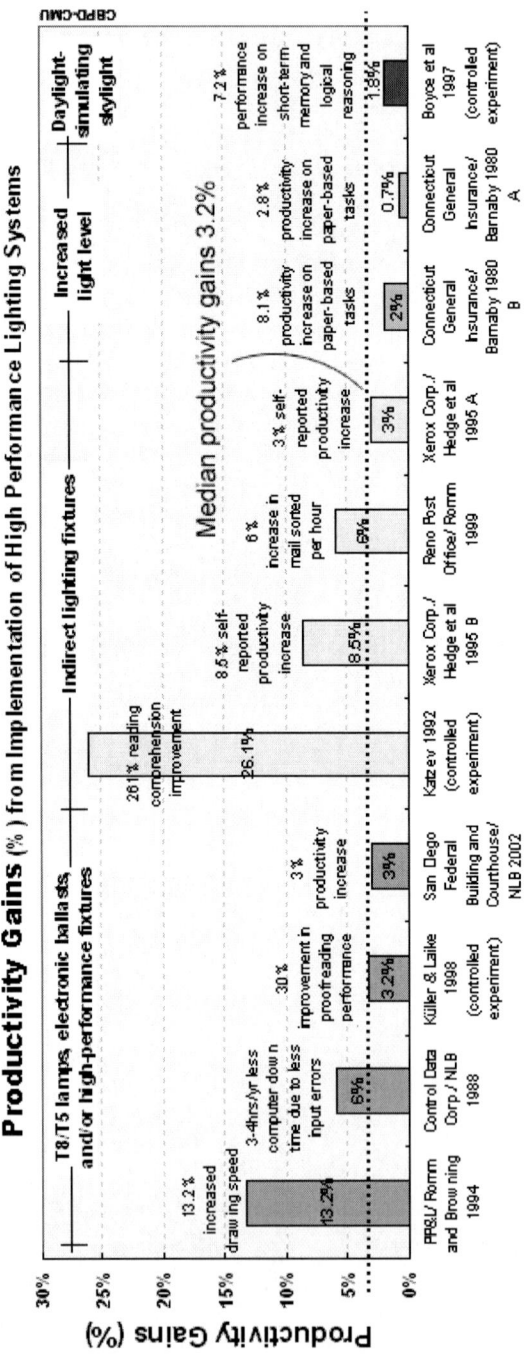

8.3.4 Innovative systems integration: under-floor air

There are a growing number of LEED™ Silver, Gold and Platinum projects that have demonstrated measurable energy benefits as well as reduced absenteeism, quicker attraction rates and better perceived health. The difficulty lies is determining which elements of the building contributed most significantly to those gains—access to the natural environment, high-performance equipment, higher quality materials and finishes or better coordination of the professional disciplines to create a more sustainable building. As we strive for innovation in buildings to ensure health and productivity, organisational and technological flexibility and environmental sustainability it will be critical to tease out the importance of quality in each building subsystem and system integration. To this end, one systems integration innovation, the use of under-floor air to ensure task air for each individual, has demonstrated life-cycle benefits.

Under-floor air systems allow conditioned air to be supplied directly to occupants, replacing general space conditioning systems from the ceiling or wall with occupant-controlled conditioning from the floor. Twelve studies have shown that under-floor air systems can pay for themselves in less than a year as a result of the average 15% HVAC energy savings, 0.5% productivity gains and 80% churn cost savings (Fig. 8.9), as well as facility management benefits. The BIDS™ case studies demonstrate that under-floor air systems yield annual energy cost savings of $30 per employee, productivity gains of $254 per employee, churn cost savings of $154 per employee, and facilities management savings of $38 per employee, for total savings of up to $486 per employee annually (Fig. 8.10). With a one-time first-cost premium of $54 per employee for new construction

FIGURE 8.9 Churn cost savings from under-floor air systems

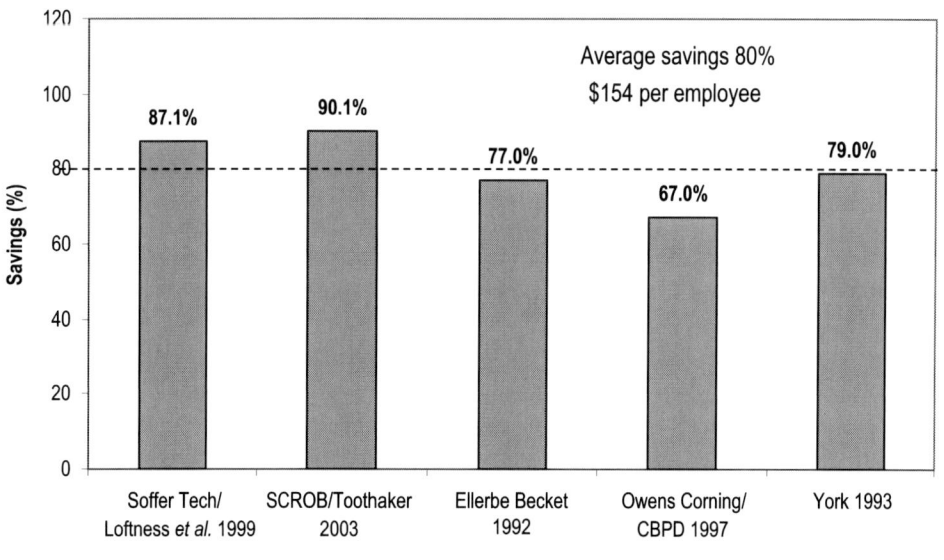

FIGURE 8.10 Costs and benefits of under-floor air systems

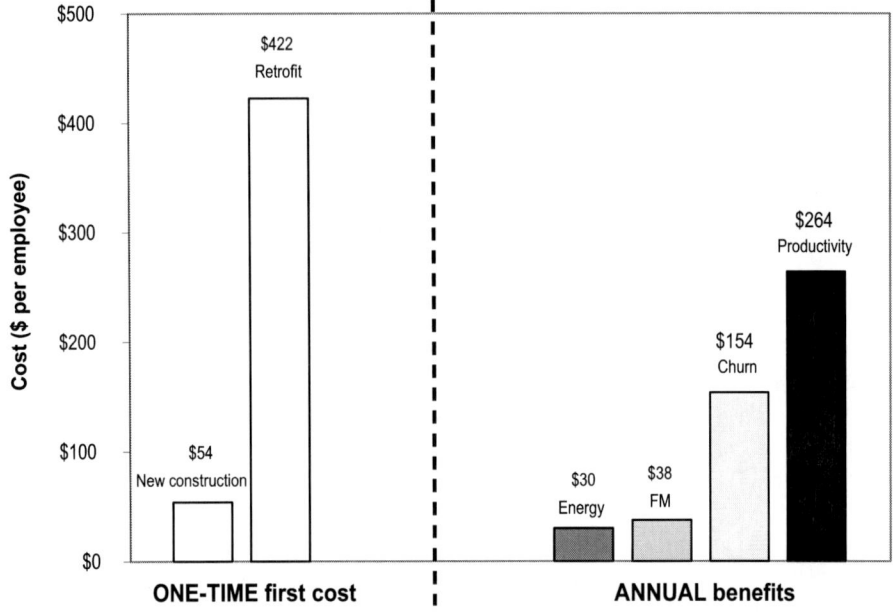

and $422 per employee to modify existing buildings, the average ROI for an investment in under-floor air systems is 900% for new buildings and 115% for retrofits.

8.3.5 Cool roofs and cool communities

Where once shading by means of massing, orientation and external and internal shading devices was integral with the aesthetics of place, the shading of buildings and communities today is a lost art. Again, first-least-cost decision-making will not support the dynamic and elegantly crafted solutions for shading that are invaluable to sustainable environments. Consequently, we must build a case for the life-cycle value of shading, landscaping and cool roof technologies, searching for energy, health, crime reduction, maintenance and other benefits to promote investments in quality built environments.

On a national level, the creation of 'cool communities' in the USA through the widespread introduction of green or white roofs, pervious paving and shade trees would yield a 10% reduction in annual cooling loads and a 5% reduction in peak cooling loads (Rosenfeld 2003). The CMU BIDS™ team has identified seven case studies indicating that cool roofs reduce annual cooling energy consumption by an average of 11%, and reduce average peak cooling demand by 14%. Given the negligible price penalty for light-coloured roofs (an additional $0.002 per square metre), cooling energy savings can pay back the initial investment in one to five years. Despite an installed cost approximately four times that found in Germany (Philippi 2006), green roof technologies are rapidly gaining favour as both a 'cool roof' strategy and a storm-water management technique in the USA. Cost savings from these benefits result in an average ROI of 7% for green roofs (Fig. 8.11).

FIGURE 8.11 Cooling energy savings from cool roofs

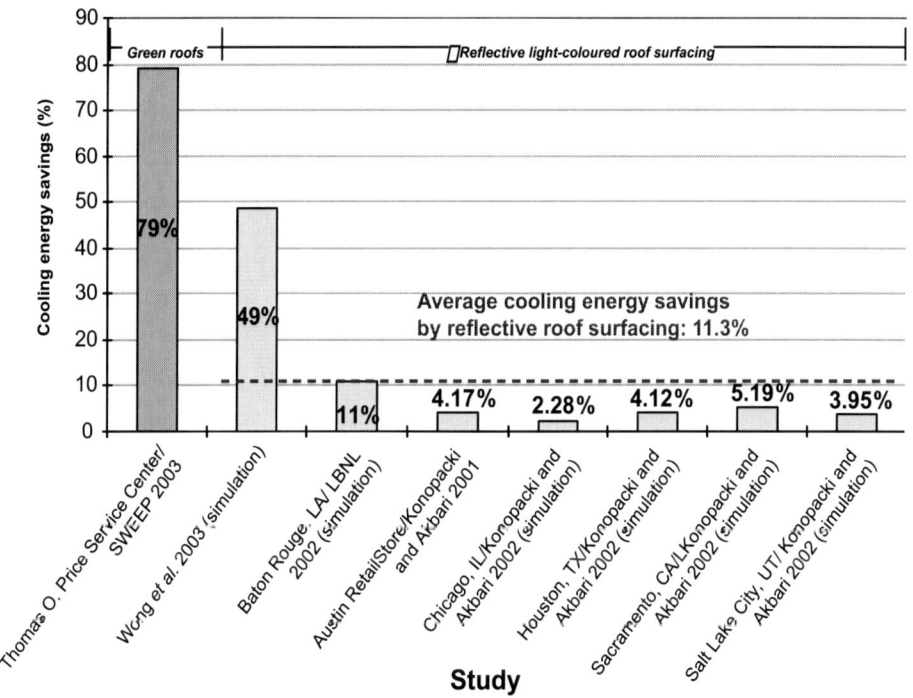

8.4 Greening the existing building: the Adaptable Workplace Lab

The BIDS™ cross-sectional findings that support daylight, natural ventilation and task lighting and air are especially valuable to the renovation of existing buildings that were once naturally conditioned. In 1999, the US General Services Administration (GSA) had a significant opportunity to create an advanced workplace laboratory on the 7th floor of the GSA headquarters building in Washington, DC. The central goal of the 10,000 square foot Adaptable Workplace Lab (AWL) was to create a collaborative and open environment that enables organisational agility. A second key goal of the project was to test how well a wide range of technologies and design features—lighting, HVAC, connectivity, interior systems and the space delivery process—support the goals of adaptability. Key performance issues included enabling staff to work remotely, managing high levels of churn, supporting work–life balance and providing for a high-quality indoor air and thermal environment. The AWL features a variety of high-performance systems and components, including raised flooring, plug-and-play (non-embedded) technologies and individual control of environmental systems, workstations and workgroups.

In 2000, a CBPD team conducted a post-occupancy evaluation of the AWL. The initial focus was on the performance of HVAC and technology systems. The raised-floor system successfully allowed for rapid changes in communications infrastructure and the HVAC ductwork to support space density and layout changes. The quality of the air was excellent, as was the daylighting. Removable floor panels allowed the space to be reconfigured quickly and at low cost, and the increased number and variety of collaboration spaces were highly valued. However, the elimination of closed offices and the downsizing of workstations to create more shared spaces (Fig. 8.12) was not popular, especially in a building with a tradition of closed or shared closed offices. The promise of individual control over temperature was never realised, as central controls took precedent over individual units. The lessons learned about spatial, thermal, visual, acoustic and air quality in open office environments, from surveys, interviews and field diagnostics, are invaluable to ongoing advances in office planning and design.

FIGURE 8.12 The Adaptable Workplace Lab, Washington, DC

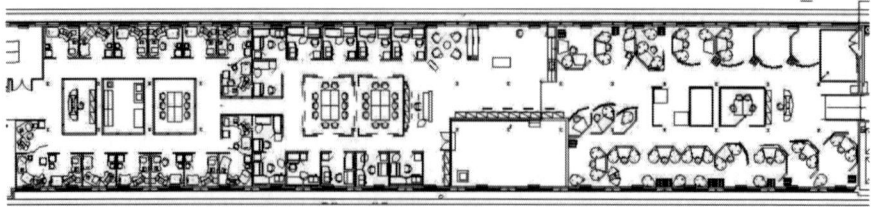

8.5 Documenting change: Workplace 20|20 and NEAT

A concerted effort is needed by the sustainable design community to capture the base-line before and after thermal, visual, acoustic, air quality and spatial quality data in the buildings they are designing or retrofitting, to supplement the findings of BIDS™ and to further encourage public and private investment in higher performance buildings. However, the task of accurately measuring the indoor environment and occupants' relationship to their physical workplace requires a commitment to field evaluation that typically lies outside the scope of the construction process. To close the gap between practice and research, in 2001 the CBPD began the development of building evaluation protocols linking environmental, technical and spatial quality to individual and organi-sational effectiveness.

Supported by the US GSA Workplace 20|20 initiative, the CBPD team developed a framework for productivity research that could be accomplished by an intensive effort in the federal sector to collect before-and-after data on key building attributes, key economic indices linked to productivity and effectiveness, and subjective and objective field measures of environmental quality. To effectively complete research results link-ing the quality of work environments to business objectives, the CBPD has been actively working to create a comprehensive 'before' profile of the GSA's innovative workplace projects in the development of multiyear before-and-after case studies. These compre-hensive profiles include subjective evaluations, objective measurements and baseline physical attributes. Field studies have been completed at 34 sites in 15 federal facilities that are due for renovation.

Integral to this process was the development of the National Environmental Assess-ment Toolkit (NEAT), a post-occupancy evaluation methodology that is designed to collect information about indoor environment conditions and user satisfaction and performance metrics at the individual workstation level. NEAT is a mobile system that captures spot and 24 hour continuous measurements of light level, brightness and contrast, air temperature and speed, radiant temperature, background noise level and indoor air-quality factors. The objective measurements are then compared with simul-taneous and long-term survey data regarding user satisfaction with environmental quality and perceived impact of indoor environment on job satisfaction and individual productivity.

The GSA case studies in facilities across the nation are providing a critical mass of measured and user satisfaction data to establish baseline conditions in federal work-places for thermal, acoustic, visual, spatial and air-quality factors. These baseline condi-tions, in combination with financial and business measures, are critical for 'proving' the impact of high-performance workplaces on 'financial' outcomes and for setting higher goals for the next generation of work environments. The data collection efforts have finally begun to reach critical mass, in terms of the number of workstations and build-ings studied, to draw conclusions about the relevance of indoor environmental and spatial quality on user comfort, satisfaction and performance. For example, field meas-urements reveal that many federal workplaces are unnecessarily cold in the summer, with energy and comfort consequences. Raising summer temperatures to levels that would support summer clothing choices for all workplace occupants will contribute to

FIGURE 8.13 Air temperature spot measurements at 624 workstations in 20 US federal office buildings nationwide (32 floors)

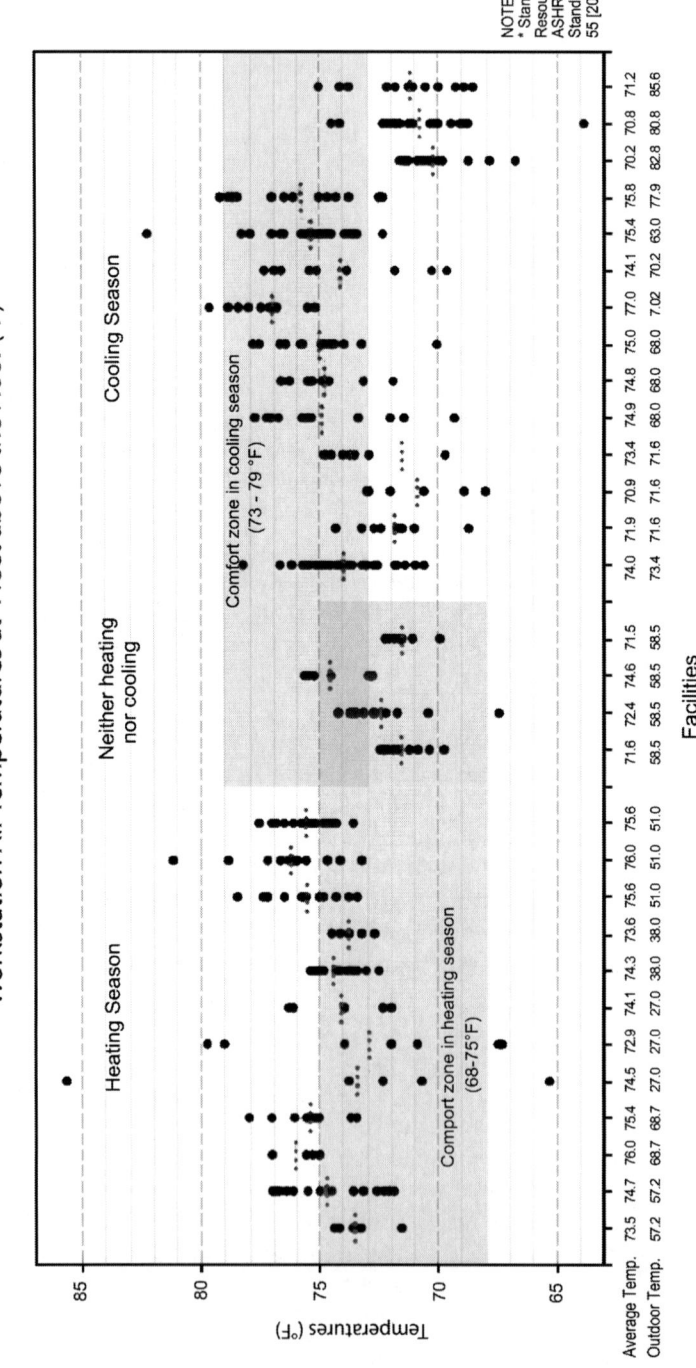

Workstation Air Temperatures at 4 feet above the Floor (°F)

increased user satisfaction and performance and measurable reductions in energy and carbon savings for federal facilities (see Fig. 8.13).

8.6 Conclusions

Investments in environmentally responsive, high-performance building solutions should no longer be limited by first-cost decision-making. Professionals should know the full cost of ownership for buildings and their impact on individual and organisational health and productivity. Clients should commit themselves to tracking building and human cost benefits over time, with a goal of understanding the life-cycle value of each material, component and system. We must close the loop on building quality and human and organisational costs and benefits over time towards achieving a sustainable built environment.

Sustainable design is a collective process whereby the built environment achieves new levels of ecological balance in new and retrofit construction, towards the long-term viability and humanisation of architecture. Focusing on environmental context, sustainable design merges the natural, minimum resource conditioning solutions of the past (daylight, solar heat and natural ventilation) with the innovative technologies of the present, into an integrated 'intelligent' system that supports individual control with expert negotiation for resource consciousness. Sustainable design rediscovers the social, environmental and technical values of pedestrian, mixed-use communities, fully using existing infrastructure, including 'main streets' and small town planning principles and recapturing indoor–outdoor relationships. Sustainable design avoids the further thinning out of land use and the dislocated placement of buildings and functions caused by single-use zoning. Sustainable design introduces benign, non-polluting materials and assemblies with lower embodied and operating energy requirements and higher durability and recyclability. Finally, sustainable design offers architecture of long-term value through 'forgiving' and modifiable building systems, achieved through life-cycle instead of least-cost investments and through timeless delight and craftsmanship (Loftness *et al.* 2005).

Decision-support tools

BIDS™ (Building Investment Decision Support) Tool, Centre for Building Performance and Diagnostics, Carnegie Mellon University; cbpd.arc.cmu.edu/bidstrial (accessed September 2010).

eBIDS (Energy Building Investment Decision Support, Centre for Building Performance and Diagnostics, Carnegie Mellon University; cbpd.arc.cmu.edu/ebids (accessed October 2009).

References

Barnaby, J.F. (1980) 'Lighting for Productivity Gains', *Lighting Design and Application* 10.2: 20-28.

Bordass, B., R. Cohen and M. Standeven (1999) 'Final Report 2: Technical Review', *Probe Strategic Review 1999* (August 1999); www.usablebuildings.co.uk/Probe/ProbePDFs/SR2.pdf (accessed September 2010).

Boyce, P.R., J.W. Beckstead, N.H. Eklund, R.W. Strobel and M.S. Rea (1997) 'Lighting the Graveyard Shift: The Influence of a Daylight-Simulating Skylight on the Task Performance and Mood of Nightshift Workers', *Lighting Research and Technology* 29.3: 105-34.

CBPD (Center for Building Performance and Diagnostics) (1997) 'Building Study of Owens Corning World Headquarters', Center for Building Performance and Diagnostics (CBPD), Carnegie Mellon University, Pittsburgh, PA.

Davis, L., L.F. Matthiessen and P. Morris (2007) 'The Cost of Green Revisited'; www.davislangdon.com/USA/Research/ResearchFinder/2007-The-Cost-of-Green-Revisited (accessed September 2010).

Ellerbe Becket (1992) 'GSA Access Floor Study', US General Services Administration, 10 September 1992.

Figueiro, M., M. Rea, R. Stevens and A. Rea (2002) 'Daylight and Productivity: A Field Study', in *Proceedings of the 2002 ACEEE Summer Study on Energy Efficiency in Buildings, Pacific Grove, CA*.

Fontoynot, M., W. Place and F. Bauman (1984) 'Impact of Electric Lighting Efficiency on the Energy Saving Potential of Daylighting from Roof Monitors', *Energy and Buildings* 6.2: 375-86.

Graudenz, G.S., C.H. Oliveira, A. Tribess, C. Mendes, Jr, M. Latorre and J. Kalil (2005) 'Association of Air Conditioning with Respiratory Symptoms in Office Workers in Tropical Climate', *Indoor Air* 15: 62-66.

Hedge, A., *et al.* (1989) 'Indoor Air Quality and Health in Two Office Buildings with Different Ventilation Systems', *Environmental International* 5: 115-28.

——, W. Sims and F. Becker (1995) 'Effects of Lensed-Indirect and Parabolic Lighting on the Satisfaction: Visual Health and Productivity of Office Workers', *Ergonomics* 38.2: 260-80.

Heschong, L., R.L. Wright and S. Okura (2002a) 'Daylighting Impacts on Retail Sales Performance', *Journal of the Illuminating Engineering Society* 31.2: 21-25.

——, R.L. Wright and S. Okura (2002b) 'Daylighting Impacts on Human Performance in Schools', *Journal of the Illuminating Engineering Society* 31.2: 101-14.

Kats, G. (2009) *Greening Our Built World: Costs, Benefits, and Strategies* (Washington, DC: Island Press).

Katzev, R. (1992) 'The Impact of Energy-efficient Office Lighting Strategies on Employee Satisfaction and Productivity', *Environment and Behavior* 24.6: 759-78.

Konopacki, S., and H. Akbari (2001) 'Measured Energy Savings and Demand Reduction from a Reflective Roof Membrane on a Large Retail Store in Austin', Lawrence Berkeley National Laboratory Report LBNL-47149, Berkeley, CA.

—— and H. Akbari. (2002) 'Energy Savings for Heat-Island Reduction Strategies in Chicago and Houston (Including Updates for Baton Rouge, Sacramento, and Salt Lake City)', Lawrence Berkeley National Laboratory Report LBNL-49638, Berkeley, CA.

Koomey, J.G. (1996) 'Trends in Carbon Emissions from US Residential and Commercial Buildings: Implications for Policy Priorities', in *Proceedings of the Climate Change Analysis Workshop, Springfield, VA, 6–7 June 1996*; LBNL-39421.

Kroeling, P. (1988) 'Health and Well-being Disorders in Air Conditioned Buildings; Comparative Investigations of the "Building Illness" Syndrome', *Energy and Buildings* 11.1–3: 277-82.

Kuller, R., and T. Laike (1998) 'The Impact of Flicker from Fluorescent Lighting on Well-being, Performance and Physiological Arousal', *Ergonomics* 41.4: 433-47.

Leaman, A. (2001) 'Productivity Improvement', *Buildings in Value* 3 (Building Use Studies Ltd).

Loftness, V., and V. Hartkopf (2002) 'Building Investment Decision Support (BIDS): Cost–Benefit Tool to Promote High Performance Components, Flexible Infrastructures and Systems Integration for Sustainable Commercial Buildings and Productive Organisations', *The Austin Papers*: 46-52; https://www.buildinggreen.com/ecommerce/cat.cfm?catid=40 (accessed September 2010).

—— *et al.* (1999) 'Sustainable Development Alternatives for Speculative Office Buildings: A Case Study of the Soffer Tech Office Building', Center for Building Performance and Diagnostics (CBPD), Carnegie Mellon University, Pittsburgh, PA.

——, V. Hartkopf, A. Aziz, S. Lee and Khee Poh Lam (2005) 'Designing a Sustainable Built Environment', in P. Matson (ed.), *Annual Review of Environment and Resources* 30 (Palo Alto, CA: Annual Reviews).

Mendell, M.J. (1991) 'Risk Factors for Work-Related Symptoms in Northern California Office Workers', unpublished doctoral dissertation, University of California.

Muhic, S., and V. Butala (2004) 'The Influence of Indoor Environment in Office Buildings on their Occupants: Expected–Unexpected', *Building and Environment* 39: 289-96.

Nicklas, M.H., and G.B. Bailey (1996a) 'Energy Performance of Daylit Schools in North Carolina'; www.innovativedesign.net/paper.htm (accessed September 2010).

—— and G.B. Bailey (1996b) 'Analysis of the Performance of Students in Daylit Schools'; www.innovativedesign.net/paper.htm (accessed September 2010).

NLB (National Lighting Bureau) (1988) 'The NLB Guide to Office Lighting and Productivity'; www.nlb.org/publications/csh federal.html (accessed September 2010).

—— (2002) 'High Benefit Lighting: Federal Building and Courthouse Save Taxpayers Money'; www.nlb.org/publications/csh_federal.html (accessed September 2010).

Pacific Gas and Electric Company (1999) 'From sunrise to sunset—this ACE is a well-lit place', Pacific Gas and Electric Daylighting Initiative; www.pge.com/003_save_energy/003c_edu_train/pec/daylight/di_pubs/1487ACE_repaginated.pdf (accessed September 2010).

Philippi, P. (2006) 'How to get Cost Reduction in Green Roof Construction', in *Proceedings of Greening Rooftops for Sustainable Communities, 11–12 May 2006, Boston, MA, USA*; www.greenroofs.org/index.php/annualconferences/buyproceedings (accessed September 2010).

Preziosi, P., S. Czerniichow, P. Gehanno and S. Hercberg (2004) 'Workplace Air-conditioning and Health Services Attendance among French Middle-aged Women: A Prospective Cohort Study', *International Journal of Epidemiology* 33.5: 1,120-23.

Robertson, A.S., K.T. Roberts, P.S. Burge and G. Raw (1990) 'The Effect of Change in Building Ventilation Category on Sickness Absence Rates and the Prevalence of Sick Building Syndrome', *Proceedings of Indoor Air* (Toronto, Canada) 90: 237-42.

Romm, J.J. (1999) *Cool Companies: How the Best Businesses Boost Profits and Productivity by Cutting Greenhouse Gas Emissions* (Washington, DC: Island Press): 87-89.

—— and W.D. Browning (1994) 'Greening the Building and the Bottom Line: Increasing Productivity through Energy-efficient Design', Rocky Mountain Institute; www.rmi.org/images/other/GDS-GBBL.pdf (accessed September 2010).

Rosenfeld, A.H. (2003) 'Improving Energy Efficiency 2–3%/year to Save Money and Avoid Global Warming', paper presented at the Sessler Symposium, Lawrence Berkeley National Laboratory, 15 March 2003; www.energy.ca.gov/commissioners/rosenfeld_docs/index.html (accessed September 2010).

——, P. McAuliffe and J. Wilson (2004) 'Energy Efficiency and Climate Change', in C. Cleveland (ed.), *Encyclopedia of Energy* (London: Academic Press, Elsevier Science); www.energy.ca.gov/commissioners/rosenfeld_docs/index.html (accessed September 2010).

Rowe, D. (2002) 'Pilot Study Report: Wilkinson Building', The University of Sydney, Sydney, Australia.

Schrum, L., and D.S. Parker (1996) 'DOE-2 VALIDATION: Daylighting Dimming and Energy Savings: The Effects of Window Orientation and Blinds', *Building Energy Simulation User News* 17.1: 8-16; gundog.lbl.gov/dirun/1701.pdf (accessed 7 June 2003).

Skov, P., O. Valbjorn and B.V. Pedersen (1990) 'Influence of Indoor Climate on the Sick Building Syndrome in an Office Environment', *Scandinavian Journal of Work Environmental Health* 16: 363-71.

Sterling, E., and T. Sterling (1983) 'The Impact of Different Ventilation Levels and Fluorescent Lighting Types on Building Illness: An Experimental Study', *Canadian Journal of Public Health* 74 (November/December 1983).

Thayer, B.M. (1995) 'Daylighting and Productivity at Lockheed', *Solar Today* 9.

Toothaker, J. (2003) 'Churn: The High Performance Green Building', *Trump Card* 57.

US Department of Energy, Office of Energy Efficiency and Renewable Energy (2003) 'Durant Road Middle School (High Performance Buildings Database Case Study)'; www.eere.energy.gov/buildings/highperformance/case_studies/overview.cfm?ProjectID=46 (accessed September 2010).

USDS (US Department of State) (2002) *US Climate Action Report: The United States of America's Third National Communication under the United Nations Framework Convention on Climate Change* (Washington, DC: USDS. May 2002).

Wong, N.H., D.K.W. Cheong, H. Yan, J. Soh, C.L. Ong and A. Sia (2003) 'The Effects of Rooftop Garden on Energy Consumption of a Commercial Building in Singapore', *Energy and Buildings* 35.4: 353-64.

York, T.R. (1993) 'Can you Afford an Intelligent Building?', *FM Journal*, September/October 1993: 22-27.

9

Consumer feedback: a helpful tool for stimulating electricity conservation? A review of experience[1]

Corinna Fischer
Institute for Applied Ecology, Freiburg, Germany

All scenarios aiming at a sustainable energy system agree that overall energy consumption must go down (for a recent example, see Greenpeace and EREC 2007). The electricity sector is an interesting field of action within this context, for various reasons. First, electricity consumption is still on the rise in OECD countries (IEA 2006). Secondly, because of production and network losses, electricity is particularly primary energy intensive as compared with other end energy forms. One kilowatt-hour (kWh) of electricity consumed means up to 3 kWh of primary energy used.

Because of the multitude of small users and the variety of electric appliances, households seem a particularly difficult target group. In Germany, for example, the household sector is the one with the fastest growing end energy consumption.[2]

Improved feedback about electricity use is one method that might stimulate households to bring down their consumption. Feedback means individual information about a household's previous consumption. Such information is badly needed: Kempton and Layne (1994) equate consuming electricity to shopping in a grocery store where no individual item has a price marking, and the consumer receives a monthly (or, in many countries, even annual) bill on an aggregate price for 'food consumption'. She or he has

---

1 See also Fischer 2008.
2 See Arbeitsgemeinschaft Energiebilanzen, at www.ag-energiebilanzen.de.

no idea how, when or by which appliances electric current was used. Nor is he or she informed whether his or her consumption is relatively high or low (which could stimulate a search for reasons), or whether it has increased or decreased (and thus, whether his or her actions had any effect).

The essential function of better feedback is to help households to learn about their consumption (and often also about possibilities for saving). Learning, in turn, provides two ingredients that, according to psychologists, are essential for action: control and motivation. (for example, see Matthies 2005). Control means steering one's actions so that they have the desired effect. It is possible only if we can identify the effect of an action. In our case, control is enhanced by recognising how much electricity is consumed by certain actions or appliances, allowing the consumer to identify those that are most wasteful. A motivation for conserving energy may be generated when one realises that consumption is higher than expected or desired or that it produces unnecessary cost or environmental problems. Motivation may also be enhanced by the competitive factor implied in certain types of feedback (people may want to be more energy-efficient than their neighbours), or by the feeling of pride that can go with recognising that one has been able to reach a certain conservation goal.

This chapter explores the potential of various feedback techniques for bringing down household electricity consumption. Its aim is to identify which features feedback must have in order to work best.

The method of choice is a review of 26 feedback cases in 11 OECD countries (the USA, Japan, Norway, Switzerland and seven EU countries), dating from 1987 onwards. The review is based on five review studies (Abrahamse *et al.* 2005; Darby 2001, 2006; IEA 2005; Roberts and Baker 2003) and nineteen original papers.

All projects address private households, with the general aim of supporting electricity conservation, though specific aims vary (see Section 9.1.3). The projects covered vary greatly in scope, actors and motivation (see the next section for more detail). They are voluntary initiatives, generally of a bottom-up type, though national energy agencies are involved in a few of them.

All in all, they are about incremental changes on the consumption side. However, the widespread implementation of some of the reviewed methods would require institutional innovation as well as innovations of varying degrees in technical infrastructure.

9.1 Case description

9.1.1 Overview

The effectiveness of feedback for electricity conservation depends on a number of design features, including the context, location, size and specific goals of the project, as well as the different features of the feedback itself, such as frequency, content, breakdown, presentation, inclusion of comparisons, and combination with additional information and other instruments. In this section, after a short review of the landscape and regime in which the feedback projects are embedded, I will systematise the projects according to these design features.

9.1.2 Case context: landscape and regime

9.1.2.1 Landscape factors: liberalised markets, climate change and energy security

This chapter deals with the socio-technical regime of electricity consumption and production. This regime is embedded in a broader cultural landscape characterised by competing trends and views. On the one hand, the logic of globalisation, privatisation and liberalised markets is a strong driver. Electricity is no longer conceived as a basic need or public good that must be provided by the state to its citizens but as a commodity on a worldwide market. It follows that most actors' activities in this field are, and must be, governed by economic logic. On the other hand, environmental discourse is becoming stronger, especially with respect to climate change. In this way, sustainability arguments are introduced into energy policy and an incentive for political intervention in the interest of climate protection is provided. A third relevant driver is the concern for energy security, fuelled by international tensions and conflicts.

9.1.2.2 Socio-technical regime: electricity production and consumption

As the cases are located in 11 different OECD countries, there are differences in socio-technical regimes that cannot be discussed here in detail but may well influence results and have to be kept in mind when evaluating the projects. This section sums up some similarities.

The socio-technical regime of electricity production and consumption in industrialised countries is usually dominated (to varying degrees) by large-scale centralised electricity generation based on the base load concept: a few large power plants run continuously, complemented by smaller, more flexible plants in peak times. Electricity supply is dominated by a few large, influential utilities (though the Nordic countries and the UK have been more successful in reducing concentration than others) (Jamasb and Pollitt 2005). In the countries covered by the studies, the electricity market is being, or has recently been liberalised, meaning that politically guaranteed and state-regulated monopolies have been abolished, state-owned companies privatised and the state has taken on the role of a market regulator, setting a framework for the market with the aim of providing a 'level playing field' for private actors. This has enabled the emergence of new players such as independent power producers (IPPs) and energy service companies (ESCOs). It has also led to companies competing for customers. The degree of competition, however, varies greatly between countries (Jamasb and Pollitt 2005).

To users, electricity is a low-interest product. Only for a few, disadvantaged social groups does it make up a substantial share of the household budget. Furthermore, consumers experience the services provided by electricity (such as light, heat, support for housework or the use of electronic media) but not electricity itself. Electricity itself remains invisible and has no emotional value.

Currently, the regime is still in a state of malleability. Not all the ramifications of the liberalisation process have come to an end. Countries are still experimenting with methods and institutions of electricity market regulation and are trying to implement climate policy measures. The EU is developing climate and energy policies, including energy efficiency policies and policies on energy information (see Section 9.3.2.2).

This regime structure and the current regime changes have mixed implications for the implementation of feedback, as will be discussed in Section 9.3.2.1.

9.1.3 Ordering the array of feedback projects

This section gives an overview of the various types of feedback reviewed. The first sub-section will focus on the overall project design, the second on specific features of the respective types of feedback.

9.1.3.1 Overall project design

9.1.3.1.1 Context

The first important finding is that there are very few real-life experiments among the reviewed projects. Half of them are research projects, trying to test the implication of a theory or theories or to fill knowledge gaps left open by earlier research. Of those, four explore consumer preferences towards feedback: two in the form of a survey (Henryson et al. 2000; Sernhed et al. 2003), one in the form of focus-group discussions (Soós and Ürge-Vorsatz 2003) and one (Egan 1999) through a combination of interviews, a survey and the evaluation of a real-life project. The others try to explore the effects of feedback. One takes the form of a laboratory experiment (McCalley and Midden 2002); the others are field tests employing a design with a control group and one or more experimental groups that are exposed to one or more types of feedback. They therefore allow for comparisons between different treatments, at least within a study.

Eleven projects are model projects, meaning that a specific type of feedback is tested 'in the field', usually in order to find out about its specific merits and its possibility for broader application.

Only two evaluations of 'real life' projects are included, one in Denmark (Karbo and Larsen 2005), and one in Norway (Wilhite and Ling 1995; Wilhite et al.1999).[3] This means that many project designs will not necessarily be fit for application in the real world, for example regarding cost efficiency or technical requirements. The lack of reported real-life projects indicates potentially severe problems with putting existing knowledge about feedback into practice. Potential reasons and remedies will be discussed in the conclusions.

---

3   The latter emerged from a research and pilot project involving the everyday billing practice of a Norwegian electric utility and has since become the basis for binding legislation. Since July 1997, regulation requires all Norwegian utilities to provide billing based on actual use, at least each quarter, and a bar-chart showing a 12-month historic self-comparison. (I thank Anita Eide for information on the legislation.)

9.1.3.1.2 Specific goals

With regard to providing feedback on electricity consumption, one may pursue different goals. To motivate and enable households to lower overall consumption is the most prominent goal, but feedback is also given for other reasons. This must be kept in mind when evaluating results, as different methods of feedback may have different degrees of success with respect to various goals.

Of the 26 projects reviewed, 23 explicitly state goals. The main reasons for giving feedback were:

● To enable and motivate households to conserve energy, or to 'stimulate ecological behaviour' (18 projects)

● To increase customer satisfaction or service (five projects, three of which were made in combination with energy conservation aims)

● To achieve load shifting or peak shaving (two projects, both in combination with energy conservation)

● To raise consumer 'consciousness' (one project)

● To explore consumer preferences, to determine what kind of feedback households would like to receive on their electricity bills (three projects)

● Or, less specifically, to test any 'effects' of improved feedback (two projects)

9.1.3.1.3 Size and location

Knowing the project size and location is important to assess the degree to which the results can be generalised. In particular, the location points to potential cultural, social or political differences to be taken into account. For example, there are indications that feedback works very differently in different social milieus (Nielsen 1993).

The sample of projects covers quite a range of different household types in terms of household size, features of the building, appliance stock, ownership, income and social status. In a number of projects, this mixing is done deliberately in order to achieve a representative sample. The broad array of locations covered in the review allows for some assessment of the generalisability of results.

With regard to project size, the situation is not as good, though. Many model projects and field experiments include no more than 10–44 households. This leads to sub-groups being very small (around 10 households) and raises questions about the significance of results. Three studies (Brandon and Lewis 1999; Haakana *et al.* 1997; McCalley and Midden 2002) include around 100–120 participants but by splitting them into several sub-groups again arrive at rather small sub-groups. Seven big field experiments with over 1,000 participants are not reported in detail (Henryson *et al.* 2000). This leaves us with only five well-documented projects with big samples for analysis: two field experiments (Nielsen 1993; Sexton *et al.* 1987) and three implementation studies (Egan 1999; Karbo and Larsen 2005; Wilhite *et al.*1999).

9.1.3.2 Types of feedback

The feedback described in the various papers differs in various aspects that are likely to be relevant to its success.

9.1.3.2.1 Frequency and duration

It can be expected that feedback is more effective the more directly after an action it is given because this will allow the consumer to make an easy connection between his or her actions and their consequences. Furthermore, persistent effects will be more likely if feedback is given over a longer time, because new habits can form during that time. In the reviewed projects, the frequency of feedback ranges from continuous to bimonthly feedback, with ten projects giving feedback more often than monthly, four projects giving it monthly and seven projects giving it less often. With respect to duration, there is a very clear-cut division: six projects last less than three months (usually 4–6 weeks)[4] and thirteen (including all of the billing projects) last at least nine months (up to one or several years).[5]

9.1.3.2.2 Content

Feedback may be given on electricity consumption alone (such as kilowatt-hour), on cost, or on the environmental impacts of consumption. Different content will probably activate different motives, values and norms, which in turn are relevant to different target groups. Some consumers may be highly motivated to protect the environment. For others, the cost argument will be decisive. In the projects reviewed, all three types of information are used, though the emphasis is on consumption and cost. Eighteen projects fed back information on consumption and cost, and three on consumption only (McCalley and Midden 2002; Mack and Hallmann 2004; Mosler and Gutscher 2004). Only two projects (Jensen 2003; Brandon and Lewis 1999; the latter only in one experimental condition) test the effects of environmental information; one (Soós and Ürge-Vorsatz 2003) discusses within focus groups the desirability of such information.

9.1.3.2.3 Breakdown

Feedback may become more informative if a breakdown (such as for specific rooms, appliances or times of the day) is provided. This is almost the only way of establishing consciousness of the relevance of individual actions. However, only five of the reviewed projects provide some breakdown, and two restrict themselves to a single appliance type—cooking appliances, in Mansouri and Newborough (1999) and in Wood and Newborough (2003), and washing machines in McCalley and Midden (2002). Sexton et al. (1987) provide a breakdown for all large appliances. Wilhite et al. (1999) test a breakdown for typical uses (lighting, heating and so on) based on interview data. Karbo and Larsen (2005) use a daily load curve, based on measured data, and an appliance-specific breakdown, based on interview data, both on request. Ueno et al. (2005, 2006) provide appliance-specific and time-specific breakdowns (daily and 10-daily load curve) on request, based on real consumption data.

---

4   The projects by Ueno et al. (2005, 2006) lasted longer but were evaluated at only one early point of time, namely, after they had been running for four or six weeks.
5   Both frequency and duration are not reported for all projects.

9.1.3.2.4 Medium and mode of presentation

It has long been clear from communication sciences and learning theory that the way information is presented is crucial for its adoption (Roberts and Baker 2003). Two basic media may be used: electronic media and written material. Electronic media are used in eight studies, taking different forms. One relatively unique approach is to install an electronic display directly at an appliance. The tool provides information about the consumption of this particular appliance (McCalley and Midden 2002; Mansouri and Newborough 1999; Wood and Newborough 2003). Also, an electronic, perhaps interactive, meter may show the total consumption of a household as well as provide additional information such as time-specific breakdown or cost (Jensen 2003; Sexton *et al.* 1987). Another approach is to use computers and the internet as interactive tools. A computer program is supplied with data that may stem from user input (for example on household size and appliance stock) and/or from metering of actual consumption data, and can provide the user, on request, with a broad range of information, such as load curves, appliance-specific breakdowns, comparisons and energy-saving tips (Brandon and Lewis 1999; Karbo and Larsen 2005; Ueno *et al.* 2005, 2006). Advantages of electronic feedback are its flexibility (being able to react to users' demands, and showing different kinds of information on request) and its ability to quickly process and present actual consumption data. Interactive tools may also stimulate users' curiosity and experimentation. However, electronic feedback may be difficult to access for users not used to electronic media, and interactive tools require more user involvement.

Written material may come on its own in the form of direct mailings, brochures, etc. This is done in four projects (Brandon and Lewis 1999; Haakana *et al.* 1997; Jensen 2003; Mack and Hallmann 2004). Another possibility, used by nine projects, is to use the electricity bill as a carrier of feedback information. This approach seems promising because it can be expected that the bill is read more carefully and raises more interest than additional material. Such efforts are described in Egan (1999), Henryson *et al.* (2000), Wilhite and Ling (1995) and Wilhite *et al.* (1999).

Equally important is the means of presentation. The projects apply numerous types of presentation, the most common being text, load curves, bar charts or pie charts (for an application-specific breakdown or comparisons in time and with other households), and horizontal lines or bell curves (for comparison with other households). As an example, Figure 9.1 depicts how differently comparisons with previous consumption periods are rendered in German electricity bills.

The comprehensibility and appeal of text or graphics is a crucial success factor for feedback. Still, most projects do not seem to reflect on these problems. The choice of a specific design is usually not discussed at all, nor are reasons given for a specific choice. Only two projects test design variations systematically (Egan 1999; Wilhite *et al.* 1999).[6]

---

6 After the completion of this chapter, a number of research projects have been conducted to deal with this issue in more depth. One major project can be found here: www.isoe.de/english/projects/intelliekon.htm (accessed September 2010).

FIGURE 9.1 Comparisons of current electricity consumption with the previous billing
period in German electricity bills

Example 1: Most common format of German electricity bills (source:
Sample bill from the public utility company in Georgsmarienhütte, 2006,
translation by the authors)

For your point of consumption, we supplied or removed the following between 01.01.06 and 31.12.06:

	Consumption	Previous year's consumption	Amount €
Electricity	2,306 kWh	1,837 kWh	437.00
		Total	437.00
		Minus installments paid to 05.Jan.07	360.00
		Your total due	77.89

Example 2: Recalculation to daily averages (source: Electricity bill from
Greenpeace Energy, 2006, translation by the authors)

Overview of consumption				Information on our gross prices		
	Total kWh	Days	Consumption/ day (kWh)	From (date)	Base price €/month	Working price ct/kWh
Current bill	1,012	372	2.7	01/01/05	7.85	18.40
Previous bill	932	357	2.6	01/01/06	7.85	18.90

Example 3: Graphical depiction (source: Bill from Badenova, translation
by the authors)

Your new monthly fee for electricity and natural gas is € 81,00.
For a more detailed explanation of the method used to determine your monthly payments as well as the payment dates, please see the following pages.

Overview of your electricity consumption		Overview of your natural gas consumption	
Current billing period	951 kWh	Current billing period	9,428 kWh
Last billing period	554 kWh	Last billing period	3,372 kWh

Source: Fischer and Duscha 2009

9.1.3.2.5 Comparisons

Comparisons are said to stimulate energy conservation, first by stimulating competition and ambition (motivational aspect) and second by making transparent whether consumption (for example in a certain period or of a certain household) is 'out of the norm'. The 'unusual' consumption is expected to activate a search for reasons and an attempt to redress the issue (problem awareness aspect) (see, for example, Mack and Hackmann 2007). There are two basic types of comparison: historic and normative. Historic comparison relates actual to prior consumption (often temperature-corrected comparison with the same period in the previous year; for example, see Fig. 9.1). Almost all reviewed studies present, or deal with, historic comparison (with the exception of Soós and Ürge-Vorsatz 2003).[7] Normative comparison compares consumption with that of other households (for example, with a national or regional average, with

---

7 For five of the seven studies reported in Henryson et al. 2000 historical feedback is not reported, but because all studies are under-reported by the authors and original sources are unavailable, this does not necessarily mean such feedback was not present.

other households in the neighbourhood or with households that are in some way similar, such as in terms of size, type of house or application stock) (for examples, see Fig. 9.2). Ten studies also take up this option.

FIGURE 9.2 **Four examples of normative comparison in electricity bills**

June electricity bill in the Smiths' neighbourhood

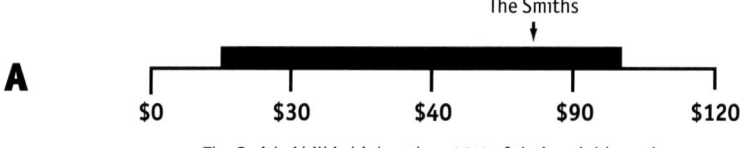

A

The Smiths' bill is higher than 90% of their neighbours'

B

The Smiths' bill is higher than 90% of their neighbours'

C

The Smiths' bill is higher than 90% of their neighbours'

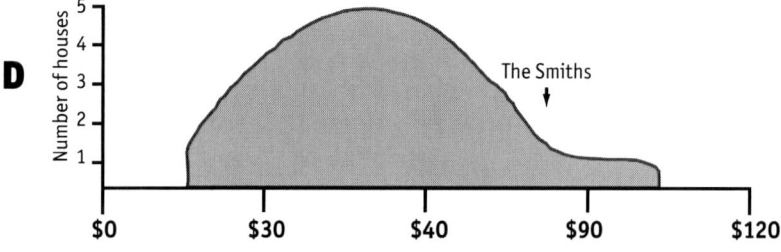

D

The Smiths' bill is higher than 90% of their neighbours'

Source: Bengtsson 1997

9.1.3.2.6 Additional information and other instruments

Feedback is very often combined with other instruments, which makes a lot of sense from a theoretical point of view (Mack and Hackmann 2007). Information on consumption will not work without the motivation to conserve, which may be provided by other instruments such as financial incentives (Nielsen 1993; Sexton *et al.* 1987), goal setting (McCalley and Midden 2002; Mosler and Gutscher 2004) or personal commitment (Mack and Hallmann 2004; Ueno *et al.* 2005, 2006.). Furthermore, feedback is of no use if households have no idea of what they can do about their consumption. This problem may be remedied by additional information on how to save energy, ideally closely connected to the appliance or situation on which feedback is given. Most projects use or explore such additional information (with the exception of Egan 1999; Jensen 2003; Sernhed *et al.* 2003; and two studies reported in Henryson *et al.* 2000).

9.2 Results

This section gives a preliminary evaluation of the results of the various feedback projects. First, it sums up the general effects of feedback, followed by a short comment on their sustainability impact. Then it draws some lessons to be learned for future feedback projects: which design features must feedback have in order to be most effective and/or attractive?

9.2.1 Main results: does feedback work?

One result, at least, seems clear: feedback stimulates energy (and specifically, electricity) savings. Savings, are, however, dependent on the target group. Three studies find no savings in low-income and/or low-consumption areas (Bittle *et al.* 1979–1980; Brandon and Lewis 1999; Nielsen 1993). Potential explanations are that there is just no more saving potential or that users realise (by comparative or historical feedback) that their consumption is already relatively low (or has been dropping), signalling that there is space for improvement on comfort. Furthermore, if feedback is designed with the main purpose of load shifting, it fails to produce overall savings. This is the case in Sexton *et al.* 1987 (see footnote 8).

9.2.2 Change in sustainability performance

9.2.2.1 Environmental improvement

As reported, feedback can reduce household energy consumption—with all the well-known positive effects on climate, emissions and resource consumption. It is very difficult to quantify the effects, however. Not all studies discuss actual savings; some concentrate on customer preferences or on satisfaction with feedback schemes. Furthermore, those who do discuss savings use very diverse reporting schemes. They vary in baseline, in time and duration of measurement and in the unit for which savings are reported (for more detail, see Fischer 2007, 2008). A very rough result is that reported savings range from 1.1% to over 20%, depending on the treatment. Usual savings are between 5% and 12%.[8]

9.2.2.2 Social and economic improvements

If feedback can reduce household electricity consumption, this has also social and economic advantages for the households. They increase their personal control over their consumption, which is in itself satisfying. And they save money that can be invested for other purposes. For utilities, however, the value of feedback is, at best, mixed. It may be a helpful tool for customer retention and/or load management and peak shaving. But whether it really provides these benefits and whether the benefits outweigh the costs is highly dependent on the individual case.

9.2.3 Learning experiences: promising design features for future feedback projects

9.2.3.1 Which types of feedback are most effective?

We not only want to know whether feedback is effective in general, but also how it must be designed to produce the best results. This question is difficult to address. Studies can be compared only with the greatest of care. First, results are not always reported quantitatively or in sufficient detail to make a comparison. Second, as mentioned above, studies use very diverse reporting schemes.

To arrive at some conclusions, I first checked 'best cases'. As such, I identified projects and experimental conditions appearing to provide the greatest savings within a group of at least roughly comparable studies. Where there was no comparable study, I identified as 'best cases' the experimental conditions providing the greatest savings within the particular study. I found that, with regard to the design features discussed above, at least two (and usually at least three) of the following features were present in the best cases:

- There were multiple options for computerised feedback available to the user (for example, the user could choose to compare consumption over various time-periods, make other comparisons or access additional information such as their environmental impact or energy-saving tips)

---

8 Information on the statistical significance of the findings is often lacking, but the sheer number of studies that report savings is a good indicator for the general effectiveness of feedback. One study (Sexton *et al.* 1987) finds no effect at all, but this is explained easily by the study design, which focuses on load shifting and combines feedback with massive price differences between peak and off-peak hours. In this case, consumers over-compensate savings in peak hours with consumption increases in off-peak times.

- There was an interactive element that engaged households in an activity, through computerised feedback or through required activities such as meter reading

- Feedback was given more often than monthly (either continuously, through daily load curves or immediately after the action)

- A detailed, appliance-specific, breakdown of energy use was provided

- Comparisons with previous periods were provided

However, results are not overly clear, because there are also less successful cases exhibiting the same features. The only feature that appears exclusively in the best cases (but not in all of them) is computerised feedback. Therefore, as a second step, I compared, for each design feature, the performance of cases that include such feedback with the performance of those that do not. That is, I grouped projects into sub-groups with high versus low frequency of feedback, long versus short duration of feedback, different content and so forth, and compared the performance of the respective groups (see for details, see Fischer 2007). The results are described in the following sub-sections.

9.2.3.1.1 Frequency
Only projects that provide feedback at least monthly are among the best performing ones. However, some of the projects that provide feedback more than monthly are also quite low performing. This indicates that frequent feedback is a necessary but not sufficient condition for best performance.

9.2.3.1.2 Duration
There is no clear indication that long-term projects provide higher (initial) savings than short-term projects. However, it seems sensible to assume that long-term projects contribute to habit formation and can therefore engender more persistent savings after treatment (which have not been reported).

9.2.3.1.3 Content
As almost all projects combine consumption and cost information, there is no basis for separating the effects of these types of information. It is possible only to look separately at the two cases in which environmental feedback is given. Environmental information proves effective in one of these cases (Jensen 2003) but not in the other (Brandon and Lewis 1999). The findings suggest at least that environmental information may be as effective as other kinds of information—depending on the target group. A sensible conclusion would be to tailor the type of information given to the potential motives and norms of the target group.

9.2.3.1.4 Breakdown
Lamentably, very few of the projects using appliance-specific breakdowns provide reliable data for judging the effectiveness of this measure. Of the three projects that do, two (McCalley and Midden 2002; Ueno *et al.* 2005) are among the most successful—a

good indication of the potential usefulness of detailed, appliance-specific data.[9] Also, appliance-specific data is appreciated by consumers (Wilhite *et al.* 1999).

9.2.3.1.5 Medium and mode of presentation

We have already seen that interactive, computerised feedback is very stimulating. Interactivity and the possibility of choice involve customers, raise their attention, and allow for tailored solutions. It is less clear, however, what exactly the presentation must look like. Surprisingly, very few studies have considered the relevance of graphic design or the formulation of text. Roberts and Baker (2003) suggest that the presentation should be simple but not simplistic, that it should not involve additional paper and that a combination of text, diagrams and tables is more effective than single-format presentations. This is a start, but there is not yet enough detail. The only two comparative studies show convincingly that household reactions to graphical designs depend very much on the exact choice of diagram or chart type, labels, scale, symbols and wording of the explanation. Designs may range from the completely unintelligible to the highly motivating (Egan 1999; Wilhite *et al.* 1999). For example, of the four graphic designs presented in Figure 9.2 (page 141), design C was apparently the most comprehensible for Delaware respondents, with 79% of them correctly answering a question about the Smiths's consumption. In contrast, only 59% of respondents interpreted design B correctly. Furthermore, 86% said that design C would motivate them to save energy, and only 70% said so for design B (Egan 1999).

A special case is the use of the bill as a medium. None of the billing projects is among those yielding the highest scores. They show quite a range of savings, from 0% (only one case) to 12%. However, billing projects have other advantages, to be discussed in Section 9.3.3.

9.2.3.1.6 Comparisons

As almost all projects use some form of historical comparison, it is worthwhile only to look separately at normative comparisons. It shows that none of the ten studies dealing with normative comparison could demonstrate an effect on consumption. A simple reason presents itself: Although it stimulates high users to conserve their energy use, it suggests to low users that things are not so bad and that they may increase their energy use a little! These effects probably tend to cancel out each other. A similar argument may hold for historical feedback: it stimulates conservation only when consumption has risen.

9.2.3.1.7 Additional information and other instruments

The theory postulates that motivational instruments (such as goal setting, the making of commitments or the provision of financial incentives) and information on 'how to' conserve must be present in order to make feedback work. The empirical evidence, though, is less clear. In many studies, feedback alone seems to work. One methodo-

---

9 It remains unclear, though, why the project by Ueno *et al.* (2006), which is a very similar project to Ueno *et al.* 2005, resulted in much less savings. Uncertainties arising from the very small sample surely play a part.

logical reason may be the small size of experimental groups; there are also reasons in the design of specific studies. A possible substantial explication is that motivation and knowledge about energy-saving possibilities is already present to some degree in participating households and can be activated by the provision of feedback. Finally, as already reported, the usefulness of information depends strongly on how it is presented and whether it is specific to the needs of the target group.

9.2.3.2 How would households prefer their feedback?

Feedback is not just a switch that may be flipped in order to elicit a certain behaviour. It is also a service provided to households in order to increase their control over their actions as well as to increase their satisfaction. Therefore, it is interesting to look at the results of studies that explored households' evaluation of different forms of feedback (such as Egan 1999; one study reported in Henryson *et al.* 2000; Sernhed *et al.* 2003; Soós and Ürge-Vorsatz 2003; Wilhite *et al.* 1999).

One unanimous finding is that households in all countries approve of receiving more detailed and more frequent feedback based on actual consumption (electricity bills in many countries come in the form of estimates). It gives them a sense of control and, if delivered with the bill, of being valued and kept informed by the utility company. Furthermore, there is usually an interest in comparisons with one's own previous consumption.

It is equally clear that households prefer information that is easy to understand and that they often find their current electricity bills difficult to understand. Easy-to-understand information includes (the list is not exhaustive):

- Feedback based on actual consumption in a given period (instead of offsets from previous periods, prepayments or estimates)

- Clear labelling and explanation of labels, acronyms and technical terms

- Clear indication of the various components of the electricity price

- Support from graphic presentations that are also clearly labelled. For the purpose of breakdowns, pie charts are preferred. For comparisons with previous periods, households like vertical bar charts. And for comparison with other households, horizontal bars or lines ranging from lowest to highest consumption are the design of choice, with the various levels of household consumption indicated as data points on the line

Other preferences vary highly between nations and, probably, cultures. One example is normative feedback. A number of studies report that consumers appreciate such feedback and think it would motivate them to conserve energy in cases where they consumed more energy than average (Haakana *et al.* 1997; Ueno *et al.* 2005, 2006; Wilhite *et al.* 1999). Others are rather reluctant with regard to comparisons (IEA 2005: 10; Sernhed *et al.* 2003). National differences are also found with respect to the graphic design of feedback. One instructive example is a comparison between Egan (1999) and Wilhite *et al.* (1999), which tested the same four graphic designs for presenting a between-household comparison in Delaware, USA, and in Norway (see Fig.9.2, page

141). The design that ranked highest in the USA, design C in Figure 9.2, was regarded as a complete failure in Norway. By some respondents, it was characterised as childish. Others found it difficult to interpret because it remained unclear whether the houses represented individual households or aggregate data.

9.2.3.3 Summary of results

To sum up, feedback has the potential to contribute to electricity conservation in households and thus to a more sustainable system of electricity production and consumption. Although we must draw our conclusions with all due care because of data restraints, there are reasons to identify some likely features for successful feedback (meaning effective both in stimulating conservation and in satisfying the needs of households). Such feedback:

- Is based on actual consumption

- Is given frequently, at least monthly (though this alone is not sufficient)

- Involves interaction and choice for households

- Involves appliance-specific breakdowns

- Is given over a long period

- May involve historical or normative comparisons (although the effect of such comparisons on consumption is not proven, they are appreciated by consumers)

- Is presented in an understandable and appealing way

Thinking about feedback from the theoretical perspective sketched very briefly above (Section 9.1.3.1), one can see the plausibility of these findings. If feedback is about learning the effects of one's actions, it must not only be understandable but be closely linked to these actions. In this case this means it must be based on actual consumption, ideally detailed by appliance, and given shortly after the action so that the consumer can construct the link between action and effect. As learning processes need time, especially when they are about 'unlearning' established habits, feedback needs to be given over a long period. Interaction and choice have a double function: they stimulate user curiosity, thereby facilitating learning, and enhance individual control over consumption by providing exactly the sort of information that is needed or wanted. Finally, historic and normative comparisons are one possibility for providing the necessary motivation to conserve energy.[10]

In particular, the first four characteristics in the list point to the advantages of electronic metering and data processing. Giving feedback with the electricity bill, by contrast, has, by design, limitations. The feedback is relatively general, is not linked to specific actions and comes relatively infrequently compared with, for example, feedback tools that are attached to specific devices.

---

10 Saving money or protecting the environment would be other possible motivations but, as explained above (Section 9.2.3.1.3), the data was not sufficient to check such effects.

However, it is important to check whether the recommendations hold for all target groups. There is probably not 'the perfect' feedback for everybody. For example, interactive feedback may be too technically challenging for some, and appropriate changes to the bill may be a suitable medium for groups that are heavily motivated by economic considerations.

9.3 Potential for diffusion and scaling up

Most of the projects analysed have been designed without a view to 'scaling up'. Their main goals were to collect information about the effects of feedback. The preconditions for widely 'scaling up' and implementing electricity consumption feedback have not yet been extensively explored. In this section, I present some considerations about potential preconditions for such scaling up. I start by identifying research gaps that need to be filled, then continue by discussing factors that can stimulate or block improvements at the regime or landscape level and end by undertaking a preliminary assessment of possibilities for scaling up.

9.3.1 Research gaps

First, before we can scale up small-scale trials, we need more information about what exactly it is that should be scaled up. The present review gives us some ideas on what effective feedback should look like. However, results are at best preliminary, and there are a number of issues that need to be resolved.

There is a lack of well-documented studies involving large sample sizes (large-N studies) that could provide reliable data on what type of feedback will best stimulate electricity conservation. Such studies should cover a representative sample of households and vary systematically the kind of feedback given, ideally only one feature of feedback at a time. For example, research is needed on how various types of motivation may be stimulated by different types of feedback, and what type of motivation is effective in which target groups.[11]

Another research gap is the lack of international comparative studies. This review shows that there may be wide cultural and national differences not only in preferences but also in the type of information that is effective in stimulating conservation. As long as comparative studies remain unavailable, one must be careful about applying results from one country to another specific national situation.

Furthermore, specific information on some countries is completely missing. In particular, for EU accession countries and for Southern Europe the effects and preferred types of feedback still remain to be investigated.

---

11 Since the completion of this chapter, some larger field tests have been brought under way, in Germany for example under the 'E-Energy' programme (www.e-energy.de/en/32.php [accessed September 2010]).

Finally, feedback projects should be based on sound consumer research (Roberts and Baker 2003) in order to test variants of graphic design and presentation and to explore the specific needs and preferences of different target groups, as carried out by Wilhite *et al.* (1999) and Egan (1999). First steps in this direction have, for example, been taken by the German Intelliekon project (www.isoe.de/english/projects/intelliekon.htm).

9.3.2 Stimulating and blocking factors

9.3.2.1 Actor constellation and technical preconditions

When promising methods of feedback have been identified with some certainty, we need to consider which regime and landscape factors may stimulate or block their diffusion.

On a regime level, the first problem is which actors may be interested in 'scaling up' electricity consumption feedback. There are only a few possible actors and none with a really strong motivation. Private, profit-oriented electric utilities running significant base-load capacity will be interested only to a limited degree. To such utilities, Demand-Side Management might be attractive if undertaken with the aim of smoothing load curves. Overall electricity conservation, however, is not interesting in situations of overcapacity or where cheap electricity is available on the market. It becomes an option only in a few cases; for example, when it can help to avoid expensive investment in a new power plant or network extension.

Improved feedback could also be conceived as an additional customer service that helps to improve a utility's competitive position. This effect is limited, however, to the degree that real competition exists and that utilities believe their customers to be interested in conservation. However, as explained in Section 9.1.2.2, this often is not the case. The same argument applies to other ESCOs or potentially interested actors. National energy agencies could also promote feedback, but these do not exist in every country and vary in institutional strength, mission and budget.

Apart from the potential actors, technical structure plays a role in the implementation of improved feedback. Many variants hinge on technical preconditions that are not always met (For details on the different types of feedback, see Section 9.2.3.1). For example, continuous electronic feedback requires 'smart' two-way metering technology. A similar argument applies to more frequent (for example, monthly) feedback, if meter reading is not to become overly expensive. (However, there could be ways out of the dilemma, such as self-reading of the meter.) Appliance-specific breakdowns would need even more sophisticated technology that is at the moment unlikely to be widely installed. Comparisons with similar households rely on adequate databases that need to be built up. Other forms of feedback, however, are less demanding. Comparisons with a previous period, presented in a graphical form, for example, should be feasible as should be the inclusion of environmental impact information or energy-saving tips. In some countries, advanced metering technologies are currently being introduced, providing a better basis for improved feedback (for example, in Denmark, by 2010 25% of all meters will be replaced by remote metering and two-way communication technology; Norway is conducting pilots with smart meters; Italy has also decided to implement them widely),[12] and in the UK the government has plans to equip every household with a smart meter by 2020 (Department of Energy and Climate Change 2009).

9.3.2.2 Political initiatives

The rather unfavourable constellation of actors and the various technical hurdles mean that political incentives will be necessary in order to spur the diffusion of feedback on electricity consumption. The European Union has taken on the challenge with its 2006 Energy Service Directive (Directive 2006/32/EC) which asks for the installation of smart metering devices (although under a number of preconditions), and its 2008 Communication on 'Addressing the Challenge of Energy Efficiency through Information and Communication Technologies' (COM/2008/0241) which aims at supporting 'smart metering and advanced visualisation' by information sharing and spreading best practice (EU 2008: 7-8).

9.3.3 Possibilities for scaling up

Even in a favourable political environment, feedback will be able to leave its niche 'market' only if, by design, it offers the possibilities of widespread implementation without incurring unreasonable cost or effort. This currently excludes a number of the more sophisticated approaches, especially those providing appliance-specific and other highly individualised information (such as those described in Ueno *et al.* 2005, 2006).

Also, approaches making heavy use of interactive electronic media, though very successful with certain target groups, may be difficult to transfer to other groups. An overly complex tool requiring much understanding and initiative from users may not be the tool of choice for households with lower education, lower technical interest (as may be the case for many elderly people) or those with less spare time.[13]

The requirement for simplicity and a low cost points to the advantages of using the electricity bill as a medium for communication of feedback. Though billing projects have, by design, limitations, they have the great advantage of being able to be implemented with generally little additional effort, especially in a situation where bills need to be reconsidered anyway. Furthermore, there is a good chance that bills are studied carefully by consumers. Another advantage is that they can be run as long-term projects, forming energy-conscious habits over time. This path should therefore be explored in more detail (for an in-depth discussion of innovative electricity bills, see Fischer and Duscha 2009).

Feedback initiatives should be integrated into a portfolio of policies that support each other in order to achieve energy conservation and efficiency. For example, chances of

---

12 I thank Anita Eide for information on Norway and Italy.

13 There are interesting technical solutions on the market that allow people to monitor consumption in some detail but at reasonable cost (for example, for a tool for room-specific monitoring, see www. dezem.de [accessed October 2009], or a microchip that can be installed in electrical appliances [idw-online.de/pages/de/news217707; accessed September 2010]). However, these are still not economic for private households or require widespread technical changes in appliances.

success can be greatly enhanced if institutions are created with the mission and capacity to explore the possibilities of feedback and to initiate its implementation (such as Ofgem in the UK [www.ofgem.gov.uk]). Also, people must be able to act on the information they receive, for example, by political instruments that improve the supply of energy-efficient appliances. And price incentives should be set in order to make conservation attractive. There are many more examples of possible components in a consistent policy package that cannot be presented here in detail (see Duscha 2008; Duscha *et al.* 2006). In the context of the EU energy efficiency policy, there are chances that such policy packages will be introduced.

However, EU policies leave ample space for member states to define which measures they deem appropriate and how stringently they will implement the measures. Therefore, it is up to national actors to push for changes, promote interest in sustainable energy consumption and introduce experiments in the provision of feedback. National energy agencies could be such actors. Where there is a lack of such an agency, or where such agencies are weak or disinterested, non-governmental organisations, research institutions, consumer advocacy groups and innovative utilities could take up the role. If they do not do so, widespread implementation of helpful feedback is unlikely to occur or be successful.

9.4 Overall conclusions

Improved feedback on electricity consumption can be a helpful a tool for customers to control better their consumption and, ultimately, to save energy. Savings in the range of 5–12% are not uncommon. An analysis of the international experience over 20 years reveals that the most successful feedback combines the following features: it is given frequently and over a long time, provides an appliance-specific breakdown, is presented in a clear and appealing way, uses computerised and interactive tools and may involve historic or normative comparisons.

Feedback stimulates incremental changes in the form of reduced electricity consumption in individual households. It does not, on its own, fundamentally change consumption patterns, market structures or regime institutions. However, it could form one of many useful building blocks in a policy package aimed at energy conservation and efficiency that should be developed in the context of implementing the EU energy efficiency policy initiatives.

References

Abrahamse, W., L. Steg, C. Vlek and T. Rothengatter (2005) 'A Review of Intervention Studies Aimed at Household Energy Conservation', *Journal of Environmental Psychology* 25.3: 273-91.

Bengtsson, K. (1997) 'Can Better Utility Bills Save Energy?', *Home Energy Magazine Online*, May/June 1997; www.homeenergy.org/archive/hem.dis.anl.gov/eehem/97/970510.html (accessed September 2010).

Bittle, R.G., R.M. Valesano and G.M. Thaler (1979–1980) 'The Effects of Daily Feedback on Residential Electricity Usage as a Function of Usage Level and Type of Feedback Information', *Journal of Environmental Systems* 9: 275–87.

Brandon, G., and A. Lewis (1999) 'Reducing Household Energy Consumption: A Qualitative and Quantitative Field Study', *Journal of Environmental Psychology* 19: 75-85.

Darby, S. (2001) Making it Obvious: Designing Feedback into Energy Consumption', in P. Bertoldi, A. Ricci and A. de Almeida (eds.), *Energy Efficiency in Household Appliances and Lighting* (Berlin: Springer): 685-96.

—— (2006) 'The Effectiveness of Feedback on Energy Consumption: A Review for Defra of the Literature on Metering, Billing, and Direct Displays'; www.eci.ox.ac.uk/research/energy/downloads/smart-metering-report.pdf (accessed September 2010).

Department of Energy and Climate Change, United Kingdom (2009) 'Towards a Smarter Future: Government Response to the Consultation on Electricity and gas Smart Metering'; www.decc.gov.uk/assets/decc/Consultations/Smart%20Metering%20for%20Electricity%20and%20 Gas/1_20091202094543_e_@@_ResponseElectricityGasConsultation.pdf (accessed September 2010).

Duscha, M., D. Seebach and B. Grießmann (2006) 'Politikinstrumente zur Effizienzsteigerung von Elektrogeräten und -anlagen in den Privathaushalten, Büros und im Kleinverbrauch' ('Policy Instruments for the Improvement of the Energy Efficiency of Electric Appliances in Private Households, Offices and Small Consumers') (Dessau-Roßlau, Germany: Umweltbundesamt [UBA]; www.umweltdaten.de/publikationen/fpdf-l/3054.pdf [accessed September 2010]).

—— (2008) 'Bausteine für eine kohärente Strategie zur Förderung der Stromeffizienz in den privaten Haushalten' ('Building Blocks for a Coherent Strategy for Supporting Electric Efficiency in Private Households'), in C. Fischer (ed.) *Strom sparen im Haushalt: Trends, Einsparpotenziale und neue Instrumente für eine nachhaltige Energiepolitik* (*Conserving Electricity in the Household: Trends, Conservation Potentials and New Instruments for a Sustainable Energy Policy*) (Munich: oekom verlag): 145-57.

Egan, C. (1999) 'Graphical Displays and Comparative Energy Information: What Do People Understand and Prefer?', in *Proceedings of the Summer Study of the European Council for an Energy Efficient Economy, 1999, No. 2–13* (ECEEE).

EU (European Union) (2008) 'Addressing the Challenge of Energy Efficiency through Information and Communication Technologies' (COM/2008/0241).

Fischer, C. (2007) 'Influencing Electricity Consumption via Consumer Feedback: A Review of Experience', in S. Attali and K. Tillerson (eds.), *Proceedings of the Summer Study of the European Council for an Energy Efficient Economy, 2007* (ECEEE): 1873-1884; www.eceee.org/conference_proceedings/eceee/2007/Panel_9/9.095/Paper (accessed September 2010).

—— (2008) 'Feedback on Household Energy Consumption: A Tool for Saving Energy?', *Energy Efficiency* 1: 79-104.

—— and M. Duscha (2009) 'Consumer Feedback through Informative Electricity Bills', in B. Praetorius, D. Bauknecht, M. Cames, C. Fischer, M. Pehnt, K. Schumacher and J. Voß (eds.), *Innovation for Sustainable Electricity Systems* (Heidelberg: Physica): 115-50.

Greenpeace and EREC (European Renewable Energy Council) (2007) 'Energy (R)Evolution: A Sustainable World Energy Outlook'; www.greenpeace.de/fileadmin/gpd/user_upload/themen/energie/energyrevolutionreport_engl.pdf (accessed September 2010).

Haakana, M., L. Sillanpää and M. Talsi (1997) 'The Effect of Feedback and Focused Advice on House-hold Energy Consumption', in *Proceedings of the Summer Study of the European Council for an Energy Efficient Economy, 1997*; www.eceee.org/conference_proceedings/eceee/1997/Panel_4/p4_6/ Paper (accessed September 2010).

Henryson, J., T. Håkansson and J. Pyrko (2000) 'Energy Efficiency in Buildings through Information: Swedish Perspective', *Energy Policy* 28: 169-80.

IEA (International Energy Agency) (2005) 'International Energy Agency Demand-side Management Programme, Task XI: Time of Use Pricing and Energy Use for Demand Management Delivery, Sub-task 1: Smaller Customer Energy Saving by End-use Monitoring and Feedback' (Report; IEA, May 2005; www.ieadsm.org/Files/Tasks/Task%20XI%20-%20Time%20of%20Use%20Pricing%20 and%20Energy%20Use%20for%20Demand%20Management%20Delivery/Reports/Task%20 XI%20Final%20Report%206%20Nov%2007.pdf [accessed September 2010]).

—— (2006) *Key World Energy Statistics*; www.iea.org/textbase/nppdf/free/2006/key2006.pdf (accessed September 2010).

Jamasb, T., and M. Pollitt (2005) 'Electricity Market Reform in the European Union: Review of Progress toward Liberalisation and Integration', Working Paper 05-003, Centre for Energy and Environmental Policy Research, Massachusetts Institute of Technology, Cambridge, MA; web.mit.edu/ceepr/ www/publications/workingpapers/2005-003.pdf (accessed September 2010).

Jensen, O.M. (2003) Visualisation Turns Down Energy Demand', in S. Attali, E. Métreau, M. Prone and K. Tillerson (eds.), *Proceedings of the Summer Study of the European Council for an Energy Efficient Economy, 2003* (ECEEE): 451-54; www.eceee.org/conference_proceedings/eceee/2003c/ Panel_2/2155jensen/Paper (accessed September 2010)..

Karbo, P., and T.F. Larsen (2005) 'Use of Online Measurement Data for Electricity Savings in Denmark', in S. Attali and K. Tillerson (eds.), *Proceedings of the Summer Study of the European Council for an Energy Efficient Economy, 2005* (ECEEE): 161-64; www.eceee.org/conference_proceedings/ eceee/2005c/Panel_1/1180karbo/Paper (accessed September 2010)..

Kempton, W., and L.L. Layne (1994) 'The Consumer's Energy Analysis Environment', *Energy Policy* 22.10: 857-66.

Mack, B., and P. Hackmann (2007) 'Stromsparendes Nutzungsverhalten erfolgreich fördern' (Promoting Energy-conserving User Behaviour Successfully), in C. Fischer (ed.), *Strom sparen im Haushalt: Mission Impossible? (Conserving Electricity in the Household: Mission Impossible?)* (Munich: oekom verlag): 108-23.

—— and S. Hallmann (2004) 'Strom sparen in Lummerlund: eine Interventionsstudie in einer Passiv-und Niedrigenergiehaussiedlun' ('Conserving Electricity in Lummerlung: An Intervention Study in a Passive and Low Energy House Residential Area'), *Umweltpsychologie* 8.1: 12-29.

Mansouri, I., and M. Newborough (1999) 'Dynamics of Energy Use in UK Households: End-use Monitoring Of Electric Cookers', in *Proceedings of the Summer Study of the European Council for an Energy Efficient Economy, 1999, No. 3–8*; www.eceee.org/conference_proceedings/eceee/1999/Panel_3/ p3_8/Paper (accessed September 2010).

Matthies, E. (2005) 'Wie können PsychologInnen ihr Wissen besser an die PraktikerIn bringen? Vorschlag eines neuen, integrativen Einflussschemas umweltgerechten Alltagshandelns' ('How Can Psychologists Improve their Outreach towards Practitioners? A Suggestion for a New, Integrative Model of Environmentally Sound Everyday Practice'), *Umweltpsychologie* 9.1: 62-81.

McCalley, L.T., and C.J.H. Midden (2002) 'Energy Conservation through Product-integrated Feedback: The Roles of Goal-Setting and Social Orientation', *Journal of Economic Psychology* 23: 589-603.

Mosler, H.-J., and H. Gutscher (2004) 'Die Förderung von Energiesparverhalten durch Kombination von instruierter Selbstverbreitung mit Interventionsinstrumenten' ('Promoting Energy Conserving Behaviour by Combining Instructed Self-diffusion with Intervention Instruments'), *Umweltpsychologie* 8.1: 50-65.

Nielsen, L. (1993) 'How to Get the Birds in the Bush into your Hand: Results from a Danish Research Project on Electricity Savings', *Energy Policy* 21: 1,133-44.

Roberts, S., and W. Baker (2003) 'Towards Effective Energy Information: Improving Consumer Feedback on Energy Consumption; a report to the Office of the Gas and Electricity Markets (Ofgem); www.cse.org.uk/pdf/pub1014.pdf (accessed September 2010).

Sernhed, K., J. Pyrko and J. Abaravicius (2003) 'Bill Me This Way! Customer Preferences Regarding Electricity Bills in Sweden', in S. Attali, E. Métreau, M. Prone and K. Tillerson (eds.), *Proceedings of the Summer Study of the European Council for an Energy Efficient Economy, 2003* (ECEEE): 1,147-50; www.eceee.org/conference_proceedings/eceee/2003c/Panel_6/6051sernhed/Paper (accessed September 2010).

Sexton, R.J., N. Brown Johnson and A. Konakayama (1987) 'Consumer Response to Continuous-display Electricity-use Monitors in a Time-of-use Pricing Experiment', *Journal of Consumer Research* 14: 55-62.

Soós, R., and D. Ürge-Vorsatz (2003) 'Turning Down Demand through Electricity Disclosure: Are Consumers Ready? A Survey of Hungarian Residences and Businesses', in S. Attali, E. Métreau, M. Prone and K. Tillerson (eds.), *Proceedings of the Summer Study of the European Council for an Energy Efficient Economy, 2003* (ECEEE): 1,261-72; www.eceee.org/conference_proceedings/eceee/2003c/Panel_6/6210soos/Paper (accessed September 2010).

Ueno, T., R. Inada, O. Saeki and K. Tsuji (2005) 'Effectiveness of Displaying Energy Consumption Data in Residential Houses: Analysis on How the Residents Respond', in S. Attali and K. Tillerson (eds.), *Proceedings of the Summer Study of the European Council for an Energy Efficient Economy, 2005* (ECEEE): 1,289-99; www.eceee.org/conference_proceedings/eceee/2005c/Panel_6/6100ueno/Paper (accessed September 2010).

——, R. Inada, O. Saeki and K. Tsuji (2006) 'Effectiveness of an Energy-consumption Information System on Energy Savings in Residential Houses Based on Monitored Data', *Applied Energy* 83.2: 166-83.

Wilhite, H., and R. Ling (1995) 'Measured Energy Savings From a More Informative Energy Bill', *Energy and Buildings* 22: 145-55.

——, A. Høivik and J.-G. Olsen (1999) 'Advances in the Use of Consumption Feedback Information in Energy Billing: The Experiences of a Norwegian Energy Utility', in *Proceedings of the Summer Study of the European Council for an Energy Efficient Economy, 1999, No. 3-2*; www.eceee.org/conference_proceedings/eceee/1999/Panel_3/p3_2/Paper (accessed September 2010).

Wood, G., and M. Newborough (2003) 'Dynamic Energy-consumption Indicators for Domestic Appliances: Environment, Behaviour and Design', *Energy and Buildings* 35.8: 821-41.

10
Lifestyle dynamics as a means toward the sustainability transition

Fritz Reusswig
Potsdam Institute for Climate Impact Research (PKI), Germany

Sylvia Lorek
Sustainable Europe Research Institute (SERI), Austria

Doris Fuchs
University of Münster, Germany

10.1 A sustainability transition to reduce the carbon footprint of modern consumer societies

Rising consumer demand puts a strain on the environment as increasing amounts of space, material and energy are needed to satisfy demand (Guinee 2002; Princen *et al.* 2002; Redclift 1996; Stern *et al.* 1997; Shove and Warde 1998). More material-intensive consumption is accompanied by increasing amounts of waste and emissions. According to our own assessment, about 19 million tons of industrial carbon dioxide (CO_2) emissions (25% of the total) can be attributed to direct lifestyle and consumption-related human activities, most of which occur in the industrialised world, but with a growing share in rapidly developing countries such as China or India (Reusswig *et al.* 2005).[1]

---

1 If indirect (induced) resource flows and emissions are included, the environmental impact of consumption is even higher (Hertwich *et al.* 2005), contributing substantially to the total 'ecological footprint' of a society (York *et al.* 2004), also known as its metabolism (Fischer-Kowalski and Amann 2001).

Despite their still (very) low level of material consumption and related emissions on a per capita basis, the total carbon footprint of these emerging economies has reached significant levels.[2] Owing to catch-up processes and globalisation effects, the dynamics and the environmental effects of modern consumer society is no longer confined to its historical 'region' of origin: the USA, Western Europe and, more recently, Japan. Economic growth, political modernisation and cultural globalisation not only lead to the overall growth of resource use and emissions but also change the internal composition of societies. Myers and Kent (2003) account for 1,059 million additional people having joined the global consumer class. This applies especially to the emerging middle classes in 'dynamic' countries such as China, India and Brazil (Bhalla *et al.* 2003; Consumers International 1997; Lange and Meier 2009; MGI 2006, 2007; Robison and Goodman 1996; Sridharan 2004; van Wessel 2004). Global studies show that the propensity to consumerism and the associated dreams and hopes—often fuelled by advertising and other global mass-media products—of the emerging consumer class fuel future production and consumption processes, especially in countries with a higher proportion of poor people (Environics 2002).

At the same time, the Fourth Assessment Report of the Intergovernmental Panel on Climate Change (IPCC 2007) has made it clear that the current trend in emissions and subsequent anthropogenic climate change requires immediate action both to enhance the adaptive capacity of societies as well as to mitigate against the causes of global warming. A widely held consensus among climate scientists—and a policy goal adopted by the European Union—states that the global community should try to prevent 'dangerous climate change' by limiting itself to a Global Mean Temperature (GMT) increase of + 2°C compared with pre-industrial levels (Schellnhuber *et al.* 2006; Walker and King 2008).[3] For a high-level emission country such as Germany such a goal would translate into emission reductions of about 80% (base year 2007) in order to meet a global environmental and equity goal in 2050).[4]

Such drastic reduction goals cannot be achieved by a few energy-saving light bulbs here and a few miles less by car there. This require a sustainability transition in modern societies, and in production and consumption systems in particular (Lebel 2005; Lebel *et al.* 2010). First, by the term 'sustainability transition' (cf. NRC 1999) we refer to a normatively influenced, yet fact-based concept of how humankind will have to evolve in order to meet its needs and wants based on intragenerational and intergenerational equity criteria and without dangerous interference with the Earth's ecosystems. The broad notion of 'sustainability science' (Kates *et al.* 2001), encompassing different research domains and social discourses (for example, in the UN system), might be

---

2 China has already passed the USA as the largest emitter of greenhouse gases. The rapid growth of overall emissions will also affect the accumulated emissions over time. Although the USA, Europe and Japan are still 'leading', this is predicted to change in the near future; it expected China will pass the USA in 2021, and India will draw level with Japan 10 years later (Botzen *et al.* 2008).

3 The measured increase in GMT is about +0.76°C; an additional +0.6°C is already 'in the pipeline', but has not materialised yet because of 'inertia' in the Earth system. The two-degree goal thus leaves us with a very limited window of opportunity for reducing emissions.

4 Per capita emissions of greenhouse gases in European countries are about 10 metric tons per year, but they amount to almost 20 metric tons in the USA.

regarded as a focal point for the type of science required for the concept and its further evolution.

Second, the term sustainability transition also refers to an ongoing social process in which an attempt is made to assess and realise the viability of this concept by building and creating sustainable systems, technologies and social practices that are more or less intentionally inspired by the idea of a change in existing more or less non-sustainable practices and structures. Here, the concept has clear links to ecological modernisation theory (Buttel 2003; Huber 2000; Spaargaren 2003).

The literature on sustainable transitions (Loorbach 2010; Kemp *et al*. 1998, 2005; Rotmans *et al*. 2001) offers at least four advantages from a social science oriented view:

● It takes a long-term perspective and addresses issues of timing

● It has a focus on complexity and the multi-level character of change

● It has a clear focus on technological change

● It explicitly addresses issues of management and governance, which offers a fresh alternative to the widespread sociological attitude of pure observation

As lifestyles and lifestyle changes have not as yet been explicitly addressed by the sustainability transition literature, this chapter aims to highlight the necessity to link technology and governance issues by taking lifestyle changes into account. Without asking questions such as 'Do we need this?' or 'How much is enough?' (Durning 1992) we will not be able to meet global sustainability goals. Following this intuition, the rest of the chapter is organised as follows. We would first like to embed consumption into a conceptual framework defined by lifestyles and how they influence wider social changes (Section 10.2). In a next step, we would like to illustrate one dimension of lifestyle dynamics empirically for the case of wind energy development in Germany (Section 10.3) and the case of product carbon footprint (Section 10.4). In Section 10.5, a more general perspective will be taken, looking at the change of the lifestyle composition of a society and the diffusion of pro-environmental attitudes and behaviour. Conclusions are presented in Section 10.6.

10.2 From consumption to lifestyle

'Consumption' means different things to different people. Scientists interested in material flow analysis, often influenced by physics, engineering or biology, refer to the use or flow of materials and energy through a biophysically defined system. The advantage of being physically explicit and environmentally revealing is here usually traded against the lack of understanding of economic processes and social embeddedness. When economists talk of consumption, they usually refer to (private household) purchases of goods and services on markets. Preferences, prices, quantities and market equilibria are the dominant focal points of the economist's attention, but physical flows and environmental impacts are usually lost. Ecological economics tries to overcome this deficit (Duchin 1998; Reisch and Røpke 2005). Ecological economists challenge basic assumptions of the neoclassic tradition (such as the neglect of external costs), and sometimes even address the post-purchase phase (Cogoy 1999). Sociologists, who have tended for some time to neglect consumption because of the production bias of their discipline, usually refer to the symbolic processes by which social actors express and reproduce social inequalities and cultural values. Some scholars focus more on the subjects of consumption, and others address more the meaning of consumption in (post) modern societies as a whole (Schaefer and Crane 2005; Zukin and Maguire 2004).

In this chapter, 'consumption' refers to the processes of preference formation (for example, via advertising and public communication) and the purchase, use and disposal of goods and services by individuals in private or corporate households (organisations, governments) in a social context. In complex ways it is linked to production processes, distribution structures and other systems of provision, and combines material and energy and social (symbolic) aspects (Lodziak 2002).

Consumption processes are embedded in lifestyles. The term 'lifestyle' is widely used in environmental (and even in sustainability) contexts. Many scholars, and even some practitioners, underline the need for changing 'our lifestyle'. But given the plurality of lifestyles—and lifestyle concepts in modern societies—who should change and in what direction? In order to avoid confusion, at least three levels of analysis can be distinguished here (see Table 10.1).

TABLE 10.1 Levels and dimensions of lifestyle

Level of analysis	Definition of lifestyle	Ontological reference	Main methodologies	Lifestyle dynamics
Micro	Individual ways and forms of everyday life	Individual	Narratives Observation Qualitative interviews	Biographical and intragenerational changes in attitudes, practices and habits
Meso	Group-specific patterns of leading and interpreting individual lives	Group	Qualitative and quantitative surveys Factor and cluster analysis	Changes in social capital and social structure
Macro	Typical behaviours and mentalities of societies	Society	Cultural studies Mentality history Macro sociology	Transition of mentalities, technologies and infrastructures

Environmental sociologists—and, of course, psychologists—have often looked at (pro-)environmental attitudes and behaviour changes (and barriers to them) at the micro level of individuals and households. A this micro level, the term 'lifestyle' refers to individual ways of leading one's everyday life, and this sociological tradition can be traced back to Max Weber and his analysis of the 'Protestant Ethic' (Weber 1904, 1905). The bulk of literature on (sustainable, pro-environmental) attitudes and behaviour can be related to this level.

At the macro level of society as a whole, 'lifestyle' refers to typical behaviours and mentalities, influenced by the network of social interaction and average living conditions (such as technological infrastructure). The often quoted 'American Way of Life' would be an example of a macro-level lifestyle type. Environmental sociologists have been looking at this overall level of social performance (for example, Uusitalo 1986) and change (for example, Reusswig *et al.* 2004).

At the meso level, the term 'lifestyle' refers to patterns of activities (usually in consumption and leisure), to associated attitudes and values and to characteristics of the social situation of groups of individuals. There is no commonly shared definition of the concept in sociology. Major debates about the methodologies to be used and the role of the lifestyle concept in comparison with class occur: following Bourdieu (1976) many sociologists tend to see the consumptive and expressive side of a lifestyle more as an expression of social class and its related habits, whereas sociologists such as Beck, Lash and Giddens (Beck *et al.* 1999) and Urry see lifestyles more as autonomous expressions of individual choices, independent of class (for a brief summary of this debate, see Tomlinson 2003).

We would thus like to define the term 'lifestyles' in an inclusive way, taking three main dimensions into account (Fig. 10.2): social structure, defining the resource endowment and constraints; performance, circumscribing the practical and expressive side of lifestyles; and preferences, catching the evaluative and motivational side (cf. Lüdtke 1989; Müller 1992)

FIGURE 10.1 Three main dimensions of lifestyles at the meso level

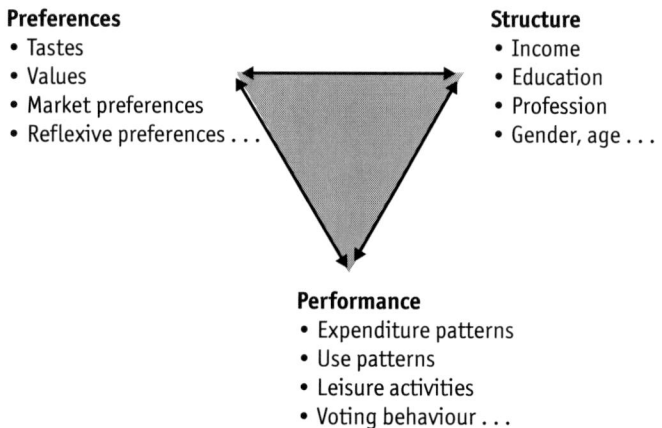

Preferences
- Tastes
- Values
- Market preferences
- Reflexive preferences . . .

Structure
- Income
- Education
- Profession
- Gender, age . . .

Performance
- Expenditure patterns
- Use patterns
- Leisure activities
- Voting behaviour . . .

Lifestyles, defined at the meso level, are group-specific forms of how individuals live their lives (performance) and interpret those lives (preferences). They imply questions of (social) identity and meaning. Lifestyles result from individual choices under social constraints (structure). Consumption is both an expression of and a resource for life-style formation and reproduction.

In modern societies, there is not one typical lifestyle, but a plurality of different life-styles, and this differentiation is important for a sustainability transition both at the descriptive and at the normative level (Reusswig 1994). Lifestyles at this social meso level have attracted less attention from environmental sociologists. However, from the areas of social structure analysis and cultural sociology an impressive body of lit-erature has emerged. In addition, commercial market and media research provides us with valuable information about market segmentation in modern societies, often speci-fied for different kinds of products and services. Studies that have looked into internal differences of modern lifestyles with regard to resource consumption and emissions reveal significant differences. Lutzenhiser and Hackett (1993), for example, found Fac-tor 4 differences between high and low CO_2 emissions in urban US households. A simi-lar study for European households detected Factor 3 differences (Weber and Perrels 2000). If 'green lifestyles' are explicitly included in the sample, differences are even larger: Christensen (1997) found Factor 8 differences between the lowest and the high-est emission families ('American Lifestyle') in Denmark.

When the analytical focus is moved from 'consumption' to 'lifestyle', consumption becomes transparent as a socially and culturally embedded process. Consumption pro-cesses and their environmental consequences can be studied in the context of social inequalities, cultural traditions and infrastructural boundary conditions.

Scholars of modern consumer society disagree about the role that consumption pro-cesses (such as preference formation, shopping, use of consumer goods, communicat-ing about them, etc.) are playing out in modern societies. Whereas some argue that consumption is more or less an appendix to the production system in modern, capital-ist, societies (for example, Fine and Leopold 1993), others highlight the structuring power of consumer preferences and consumer culture over production (for example, Campbell 1987).[5]

There is evidence for both lines of thought, and we do not see any reason why environ-mental sociology should not accept complexity and multi-causal relations. The impor-tant point is the specification of the routes of influence—in their nature and, if possible, their strength. If the conceptual focus is moved from consumption to lifestyles—and if the material and environmental aspect is not lost—this multi-causal and socially con-nected character of modern consumption processes gains greater visibility. In addition, the dynamics of consumption processes is easier to conceive. In order to elaborate this point, we would like to illustrate the complex causal relations between social institu-tions, technology and lifestyles in a simplified form (Fig. 10.2).

---

5 For an overview, see Zukin and Maguire 2004. This debate has clear consequences for environmen-tal sociology in addressing consumption issues: if the production advocates are right, every attempt to achieve more sustainable consumption via the consumption side is futile.

FIGURE 10.2 A simple model of the role of lifestyles in the context of social institutions and values and in the context of products, technologies and systems of provision; arrows show the direction of influence (or 'relations', the numbers referring to the order of relations, described in text), the arrow pointing towards the item that is being influenced and away from the item that is the influence

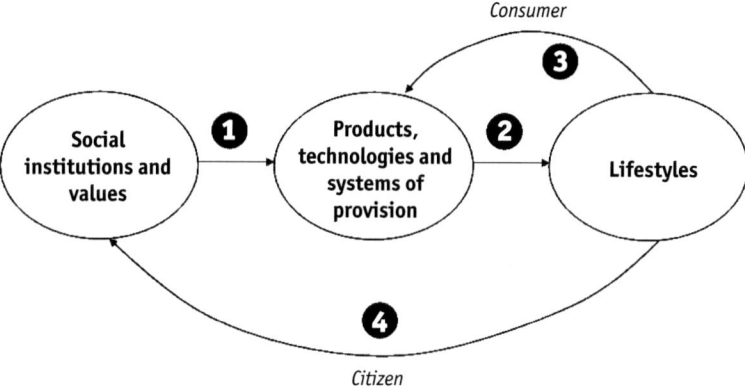

We distinguish three basic elements:

- Social institutions and values characterise the major driving forces and determinants at the social macro level

- Products, technologies and systems of provision refer to the physical 'fabric' of a society, including its organisational structure

- Lifestyles refer to individual ways of leading and interpreting one's life—be they defined purely individualistically or as group-specific patterns

The forward-facing arrows indicate two things: first, products, technologies and systems of provision are influenced by social institutions and values (Relation 1), and, second, they in turn influence (enable, constrain) the behaviour (practices, routines) and, to some degree, the attitudes of (individual) consumers (Relation 2). Relation 1 can be illustrated by studies that demonstrate how product, technology and systems development are by no means autonomous processes steered by engineers, business organisations or municipal organisations but are instead heavily influenced by underlying social institutions and their guiding problems, interest, and values. Cases in point here are research streams such as national systems of innovation (Lundvall 1995; Porter 1998), varieties of capitalism (Hall and Soskice 2001) and history of technology studies (Hughes 1983; Stier 1999).

Relation 2 has been repeatedly highlighted by environmental sociologists (sometimes even psychologists) hinting at the limits of individual (consumer) behavioural changes as drivers of a sustainability transition. Diekmann and Preisendörfer (2003) stress that environmental sociologists should take the costs of individual behaviours into account, and they highlight how social and material infrastructures influence cost structures.

Spaargaren (2003) has pointed out how closely individual lifestyle and consumption changes are linked to (and constrained by) systems of provision. His work is influenced by Cowan's (1983) analysis of the effect of infrastructural systems on the labour of private households, especially women. Shove and Warde (1998) have, in much the same spirit, highlighted the crucial interplay between systems of provision and the routines and standard expectations of everyday life.

With regard to short-term changes in consumption behaviour it might be tolerable to neglect the structural effects indicated by the Relations 1 and 2 in our simple conceptual model. With regard to long-term and system-wide changes, however, environmental social scientists more or less involuntarily promote naïve, if not ideological, research agendas by doing so. Given the complexity of social conditions for individual consumption, there are clear limits for burdening the individual consumer with the whole array of system-wide changes. Other actors, such as the business sector, civil society and governments, clearly have to play their roles as well (Jackson 2004, 2005). A closer look at the field of consumer policy reveals clear deficits and disincentives, often being much more influential on individual consumer behaviour than the widely accepted rhetoric of 'sustainable consumption' (Fuchs and Lorek 2005).

It would be naïve, if not ideological in almost the same manner, however, if environmental scientists and activists were to confine themselves exclusively to these structural effects on individual action. The rich body of literature on social change reveals a dialectical relation between individual action and social structure (Hernes 1976). For social institutions and for technologies Giddens's structuration theory (Giddens 1986) is holding as well: structures enable and constrain individual action, but individual action influences social structures. Two basic ways of individual action influencing structure should be distinguished when thinking about consumption and lifestyles: market preferences, expressed by individuals and households as consumers (Relation 3), and reflexive and policy preferences expressed by individuals as citizens (Relation 4). This double role has been illustrated by research findings on ethical consumption, consumer policy and corporate dialogues with consumer–citizens (Carrigan *et al.* 2004; Cohen *et al.* 2005; Crocker and Linden 1998; Doubleday 2004; Manoochehri 2002; Stevenson 2002).

Ulrich Beck has termed the consumer 'a sleeping giant' (Beck 1996). His or her giant-like abilities become visible only if we take both the consumer and the citizen into account. For reasons of brevity we would like to confine the influence of the consumer to products, technologies and systems of provision, whereas the citizen influences mainly the boundary conditions, that is, institutions and values. In the next sections we elaborate on the influence of citizens and consumers on technologies and social institutions. Our purpose is not to neglect the influence of technologies and social institutions on lifestyles, but to highlight how lifestyles and lifestyle changes contribute to overall social change—including a sustainability transition. Two examples will be discussed in more detail: the rise of wind energy in Germany (Section 10.3) and the introduction of product carbon footprints (Section 10.4). In both cases the interplay between consumption and lifestyle changes, technological change and institutional dynamics (including politics) is essential. In Section 10.5 we will try to embed these changes into the wider context of social change in general.

10.3 The case of the success of wind energy in Germany

The necessity to reduce the carbon footprint of the German economy was translated early on into debates about its future energy mix, including timing and costs. Most experts agree that renewable energy carriers (wind, solar, biomass, etc.) will have to play a central role. Questions are more in terms of 'how?', 'when?' and 'at what cost?'. Since the turn of the century it has become evident that we might well witness the initial phase of the required energy transition, with use of renewable energy growing rapidly—but from a rather low absolute level. The German wind energy sector displays remarkable growth rates. Since the 1990s it has outperformed the early leaders such as the wind energy sectors in Denmark and the USA. One of the major reasons for Germany being so successful in implementing wind energy is political. Two federal legislation acts (the Feed-In Law in 1991, and the Renewable Energy Law in 2000) have spurred growth substantially. International comparisons show how the German model of guaranteed tariffs, for a limited period subsidised by electricity consumers, ensures investment security for wind energy providers and has proved its superiority to the quota model favoured by countries such as the UK or Italy (Bechberger and Reiche 2005, 2006).[6]

However, the legislative aspect is only part of the success story, and it has to be seen in a wider context—a context of a socio-ecological transformation of German society, in which lifestyle dynamics play a constitutive role. It was the interplay between the political system and the energy economy on the one hand and lifestyle dynamics on the other that created what might be termed the 'take-off phase' of the German wind energy sector. In the lifestyle dynamics domain, both the 'consumer' and the 'citizen' have been important.

Looking back to the 1970s, the German economy was based mainly on fossil fuels plus a rising share of nuclear power. The energy crises of the 1970s led to some governmental research in renewable energy systems, but the main effort was to diversify the geographical distribution of oil resources and to construct many more nuclear power plants. At the end of the 1960s, the Federal Republic of Germany—as many other countries at that time—experienced a political shift to the left, the growth of left-wing, more or less radical, political parties and groups and the rise of an 'alternative milieu'—that is, a small group of mostly young and well-educated people striving for a more 'natural' way of life, including private consumption patterns and 'green' forms of production and living together. These 'eco-pioneers' were strictly opposed to big industry and government, often living in social (and geographical) niches compared with mainstream soci-

---

6 It is worth noting that wind energy alone will not provide sufficient electricity for the economy of a highly industrialised country. Other renewables will have to be used, too. Photovoltaic and solar thermal systems have a smaller market share but show impressive growth rates in Germany, each covered by the Renewable Energy Law. The future of the transportation sector is open, as many options are open (electric cars, use of fuel cells, hybrid cars, and use of biofuels and methanol). For a transitionary period, even the option of carbon capturing and sequestration (CCS) is a climate-neutral possibility, buying time for renewable energies to gain momentum. The future of wind energy will very much depend on the development of offshore facilities, which has just begun.

ety. As the federal government propagated nuclear power as a solution to the energy crises of the 1970s, these groups fought government policies—mostly peacefully, but time and again quite violently.

The critical, catalysing event for this whole constellation was the Chernobyl accident in 1986, which substantially de-legitimised nuclear power in Germany.[7] Many proponents of the German environmental social movement felt the urgent need to do something positive and constructive, and thus the citizen's wind energy movement was born (Byzio et al. 2002). Groups of engaged citizens, organised under the umbrella of cooperative societies under German public law (allowing for risk sharing), importing Danish wind turbines to be erected in their backyards. If we take Max Weber's typology of social action into account, their activities were driven by value rationality (political goals, idealism) and emotional rationality (fear of nuclear disaster). In Northern Germany in particular, where wind conditions are favourable, this citizen-based wind energy movement gained momentum during the late 1980s, including the emergence of expertise, small businesses and pressure groups.

It is important to observe how the federal government followed a different route. Under public pressure with regard to nuclear power, research and development expenditure on renewable energy sources grew slightly during the 1980s. Together with some large energy providers, a research and test facility was built in Northern Germany in 1983, the GROWIAN project.[8] The administrative and scientific preparation of a large windmill test facility had run from 1973 to 1979, initiated by the first oil crisis. Scientific wind turbine experts were able to influence the process by allowing the assumption that basic problems had already been solved, and by advising the government to go for a big technological solution. GROWIAN's dimensions (with a hub height of 100 m; rotor blade diameter of 100 m, and generating up to 3,000 MW) were not equalled by commercial wind turbines until the beginning of the 21st century. The coalition of government officials, wind energy experts and big business representatives clearly wanted to realise a 'big solution', dwarfing contemporary Danish and US developments. However, this 'big leap forward' failed; GROWIAN ran only for a few hours in total.

At the same time, the citizen's wind energy movement followed a technologically more 'conservative' pathway, using well-established and low-risk technologies and engaging in a gradual learning-by-doing process. Despite their far-reaching energy policy goals, their everyday practice was governed by a moderate, stepwise approach,

---

7 By coincidence, 1986 was also the year when the public debate on climate change gained momentum in Germany (Weingart et al. 2000). As nuclear power generation is associated with much less CO_2 emissions than coal, oil or gas powered plants, the rise of climate change as a major issue in the environmental discourse could easily have led to a strengthening of the pro-nuclear option—a route that many nuclear power advocates clearly intended to go. However, this was not the case: the majority of the German public (and, in particular, the 'alternative milieu' of that time) was concerned about climate change (in the media often termed 'climate disaster', Klimakatastrophe), but at the same time remained deeply sceptical about the risks of nuclear power. For German environmental sociologist Ulrich Beck the year 1986 was crucial as well: the first edition of his book, Risk Society, was published immediately after the Chernobyl event, offering unprecedented public resonance to the work of a sociologist (for an English translation, see Beck 1992).

8 GROWIAN is an abbreviation of Große Windenergie-Anlage (large wind energy facility).

facilitated by economic and social constraints, such as the availability of financial means, the need to pool resources with people with a similar lifestyle, the need to build up expertise, problems with local authorities and so on. This resulted in a small-scale niche of a politically inspired production–consumption system.

The year 1991 was crucial for German wind energy. The European Union had started to think actively about the liberalisation of energy markets, and oil prices were very low. The wind energy lobby, which had emerged by that time, used the window of opportunity and teamed up with the German hydropower lobby in pushing the federal government for a feed-in tariff system. Despite some opposition from the big energy providers (which were engaged in 'swallowing' the East German energy system after reunification in 1990), the law passed under a conservative–liberal government. It provided economic security for wind energy providers, reducing their investment risks, and attracted new groups of consumers and investors. Owing to the changed incentive structure, other social groups with different motivational and attitudinal backgrounds became interested in wind energy. The 'necessity' for politically motivated idealism, so indispensable during the pioneering phase, vanished. It gave way to a much more pragmatic attitude, so that even profit-seeking behaviour entered the field. In Weber's terms, this was a shift from value rationality and emotional rationality to purpose-oriented rationality—and, after time, even new forms of traditional rationality.

The organisational structure in the production–consumption system of energy provision changed accordingly. In the beginning, a rather informal design was chosen (in the form of cooperative societies), suited for networks of people with relatively strong ties, a low degree of specialisation and open to idealism and private engagement; this form was also ideal for risk pooling and allowing members to engage in only a minimal degree of formal communication with the outside world (mainly public authorities). The governance principle can be seen as representing a network or solidarity. Today, after the introduction of two major legislative items, the market has taken over. Many of the original citizens' wind power organisations still exist. However, the majority of the capacity growth since the 1990s has been achieved by medium-sized and large joint-stock companies. Shareholders are anonymous, and the degree of specialisation is high. Internally, a more formal hierarchy is in place.

True, the legal boundary conditions for wind energy generation in Germany after 1991 were in favour of growth, fuelled by a 'normalisation of ecology' (Brand *et al.* 1997), in place of the idealism of the early years:

> Legitimacy and visions are shaped in a process of cumulative causation where institutional change, market formation, entry of firms (and other organisations) and the formation and strengthening of advocacy coalitions are the constituent parts. At the heart of that process lies the battle over the regulatory framework (Jacobsson and Lauber 2006: 272).

However, the citizens' wind energy movement of the 1980s, a clear descendent of the environmental movement, situated in the alternative milieu, was crucial for the success story of German wind energy in at least four ways.

- By creating a domestic market for small wind energy systems, the movement helped to provide demand and experience and to reduce costs

- By helping to constitute a small (domestic) industry, the movement contributed to the formation of an advocacy coalition (including political party members and scientists) that was able to actively influence the legislative process

- By helping to create human capital around the construction and maintenance of wind turbines, the movement contributed to the nurturing of professional nuclei for further development in universities, the business sector and administratively

- By combining far-reaching energy policy goals with a good deal of pragmatism and the ability to form coalitions with otherwise opposing groups, the movement paved the way for a more encompassing social consensus on renewable energy as a necessary and viable option for Germany

FIGURE 10.3 A short history of the German wind energy system in the interacting domains of (a) the economy and politics and (b) civil society and lifestyles

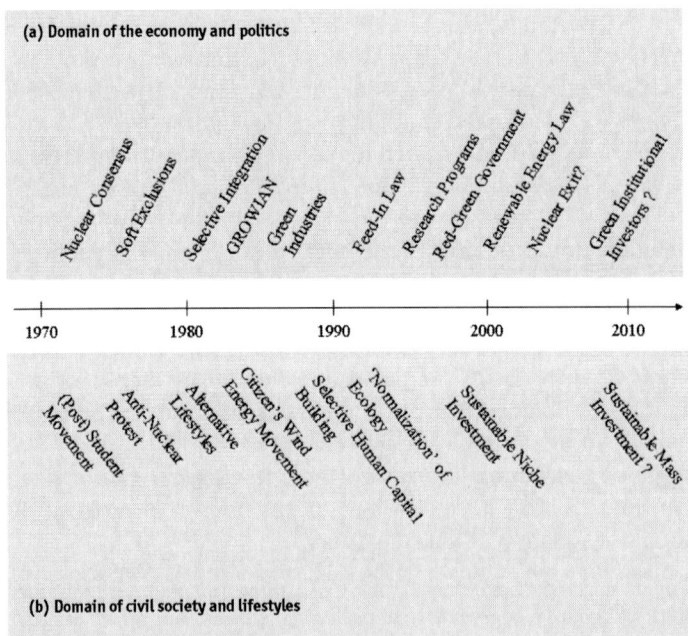

GROWIAN = Große Windenergie-Anlage (large wind energy facility)

If we look over the process from a bird's eye view (see Fig. 10.3), we observe not only phase shifts—from nuclear consensus and the (soft) exclusion of alternatives in the 1970s to the promotion of renewable energy and the nuclear exit option in the 2000s—but also changes at the lifestyle and civil society level. The important point, however, is that these two domains do not operate in isolation but are connected and influence each other. Relations 1 and 2 in Figure 10.2 tell one important part of the story: overall, social values and political decisions shape technologies and infrastructures, and technologies and infrastructures influence our consumption patterns and lifestyles. However, Relations 3 and 4 hold as well: consumers shift demand patterns, creating the need for technological changes, and citizens influence the value structure and the political process of a society, which in turn affects the technological structure. When, for example, the red–green government (coalition of Social Democrats and the Green Party) came to power in 1998, the legal boundary conditions for renewable energy had been additionally improved by the Renewable Energy Law (EEG: Erneuerbare Energie Gesetz). The point we want to make in this chapter is not to say that the citizen's wind energy movement was 'causing' Germany's international success in wind energy performance. We simply want to illustrate a multi-causal network of social and technological change, with consumers and citizens being active parts, not merely passive recipients of a sustainability transition.

In order to become really sustainable, the world's energy system has to change dramatically. The issue of energy poverty has not been touched on in this chapter so far, but should be mentioned. Given the relatively small size of European societies, the question of future energy sustainability will be answered mainly elsewhere, in China or India, for example. In these countries growth rates for renewable energy are impressive (as are other growth rates). And, in part, these rates are triggered by a network of interactions, including technology exports and international environmental agreements. However, we must not forget that consumers and citizens also play a role in these countries as well (Consumers International 1997). Membership in civil society organisations oriented towards environmental issues is growing, and it grows primarily in middle-income countries (Anheier *et al.* 2001).

There is no reason to be overly optimistic, but there is reason to go beyond a too narrowly defined image of consumption and the consumer, leaving the citizen aside. And the socioeconomic processes that are leading to a growing number of consumers in China and India also contribute—although not automatically—to higher expectations regarding environmental quality and political participation.

10.4 Product carbon footprints as a new tool for systemic innovation in producer–consumer networks

About 40% of total greenhouse gas emissions in industrialised societies arise from consumption processes and consumer goods, including food and beverages. Given the complexity of today's production–consumption networks, often organised across the globe, the reduction of the carbon footprint of consumption cannot be achieved exclusively by consumers (private households). System-wide solutions and new institutional settings (including incentives and regulations) are necessary. One key aspect, at least from a consumer point of view, is the lack of transparency. Both national and global polls on climate change perception show a widespread concern about global warming, combined with a (somewhat reduced) awareness that individual action is necessary. However, owing to a lack of (credible) information on the carbon footprint of consumer goods and consumption processes, people cannot translate their concerns as citizens into everyday choices as consumers. Assessments of greenhouse gas emissions across the total life-cycle of products exist, but they are rare, methodologically unstandard-ised and lack consistent public communication. Single aspects, such as 'food miles', arise and dominate debates, whereas other aspects (such as agricultural inputs to food products, or energy savings that can be achieved by use of insulation material) are widely neglected or circulated only occasionally. As a result, consumers—and produc-ers, although to a lesser degree—remain confused.

Given the necessities of climate policy, the world community cannot afford con-cerned, but confused, and thus 'disabled', consumers. One important step to overcome confusion and inaction has been taken by The Carbon Trust in the UK, which intro-duced a carbon label for consumer goods (The Carbon Trust 2007). Crisps (potato chips) and hair shampoo have been among the first products labelled, but the largest retailer in the UK, Tesco, has announced that it will label all its products on the shelf in the coming years. The debate on carbon labels has started all over Europe (cf. Migros in Switzerland or Leclerq in France). In Germany, a consortium of research and policy organisations (including the Öko Institute, Thema1, WWF Germany and the Potsdam Institute for Climate Impact Research) has initiated a pilot project with 10 large compa-nies (including Tchibo, dm, Deutsche Telekom and Rewe Group).[9] The main goal of this pilot project is to develop a common and scientifically sound methodology for product carbon footprints (PCFs) and to strive for international standardisation.

On average, every German consumer is responsible for about 11 tonnes of CO_2 equiv-alent per year; 4 tonnes of this (or 36%) are related to food, drinks and other consumer goods. This is more than the building-related and travel-related emissions. PCFs could be a key instrument to raise consumer awareness and to create carbon transparency in the domain of consumer goods, which currently is either lacking or distorted by single-issue foci such as an exclusive focus on food miles.

Given the global character of many value chains in today's economy, the shift towards continuous reductions in CO_2 emissions at one point affects all others, including extrac-tion, manufacturing, transportation and waste disposal.

It is clear that PCFs alone will not 'pave the way' to a low-carbon economy. However, PCFs could become a core element of such a structural shift. From an innovation-ori-ented point of view, there are some very interesting aspects to PCFs:

- PCFs are situated at the interface of production and consumption. They reflect the total value chain related emissions and communicate them to the final consumer. A CO_2 label will enable consumers to translate their concerns as citizens into consumption choices; at the same time they allow consumers to send a signal back to retailers and producers

9 See pcf-projekt.de/main/news/?lang=en (accessed October 2009). This page also provides infor-mation about the state of the international carbon footprint debate.

- PCFs make sense only if they are part of a consistent CO_2 reduction strategy within firms across the value chain. They may complement the instruments of measuring corporate carbon footprints, carbon offsetting and emissions trading, which are also evolving in the market

- PCFs will help to spur innovation processes across the value chain—both technologically (such as in the development of new refrigerators, more efficient engines or less carbon-intensive materials) and organisationally (such as the creation of energy round tables, more efficient logistics or greater stakeholder involvement). They can help to translate a given innovation potential into real innovation processes

Interestingly, it has been the private sector—with science organisations and environmental non-governmental organisations (NGOs)—that has set the pace for PCFs in Europe. Despite the fact that many firms have initially been reluctant to add another label to their products, the international dynamics of the climate and the debate on CO_2 labelling has convinced many to move forward. Given the significant recent public attention to climate change, it is very tempting for a corporation to aim to demonstrate proactive engagement—even if this claim is not backed by real improvements. This risk of 'greenwashing' can be reduced only by the use of transparent procedures, using scientifically sound and widely accepted methodologies, and by independent certification. These are the criteria that the German PCF Pilot Project follows. National governments as well as the European Commission are also considering CO_2 labels, and it remains open whether we will see a harmonised, EU-wide, solution instituted by government or whether it will be a privately driven label. In any case, PCFs and CO_2 labels provide a very promising tool in powering innovation processes towards the low-carbon economy we urgently need.

10.5 Lifestyle changes, social diffusion and the *Gestaltwandel* of environmental issues

If we move away from wind energy and PCFs we have to embed technological and consumption changes into the wider picture of social change in general. For this purpose, we will come back to the meso level of lifestyles as group-specific forms of leading and interpreting life, as sketched briefly in Section 10.2. Our goal is not only to flesh out the somewhat sketchy outline of lifestyle changes in Germany given in the previous sections but also to correct the notion of social diffusion, based on the concept of lifestyle dynamics. We will argue that the social diffusion of pro-environmental attitudes and behaviours is not adequately captured by assuming their stepwise entry in social groups other than the initial pioneers, but goes along with changes in the meaning and content of pro-environmental attitudes and behaviours themselves.

In order to grasp the meaning of this term, we have to consider changes in the structure of social milieus, as analysed by the German–French market research institute,

SinusSociovision.[10] Individual lifestyles are situated in larger social groups—social milieus—which make up the social fabric of societies.[11] Social change not only refers to a changing social landscape, with particular groups emerging, growing or declining as time goes by; it also affects the framing of issues, that is, the overall interpretation and evaluation of an issue, as well as the concrete 'tokens' (such as activities, associations and 'things') that social groups associate with a particular issue. These changes may comprise fashion, but go far beyond it, as they are linked with social status dynamics and value systems. This also holds for the issue 'environment' (or 'ecology').[12]

Wippermann (2005) has looked at the diffusion of 'ecology' (attitudes, behaviours) across German society since the early 1980s by applying the Sinus milieu approach. Starting with the 'alternative milieu', the 'breeding ground' for the environmental social movement and the Green Party during the late 1970s and early 1980s, concern for the environment made its way through the upper segments of German society (that is, through the higher status groups). A closer look shows, however, that not all higher status groups are pro-environmental, and that not all lower status groups reject the issue.

But this is not our main point. More importantly, the whole social landscape is changing, a major change being that the hardcore of socially progressive environmentalists, accounting for about 2% of the German adult population in 1982, has vanished from the map since the early 1990s. It has been in part replaced by a modernised follow-up group, aesthetically and politically less visible and radical but still strongly supporting post-materialist values and life goals. Leading an alternative way of life, engaging in environmental issues or being an ecological avant-garde is no longer the vision for the members of this post-materialist milieu. Ecology has lost its radical image. With Max Weber one might speak of a transformation of charisma to everyday life—other scholars have described this process as the 'normalisation' of ecology (Brand *et al.* 1997). One should notice, however, that post-materialism is no longer the 'spearhead' of value development in Germany. Since the 1990s, post-materialism has been 'outperformed' by issues such as 'experimentalism', hedonistic orientations and preferences for 'value hybrids'. Sinus Institute[13] approaches groups situated further 'right', such as 'modern performers' or 'experimentalists', such as the value avant-garde of German society, taking the lead in many consumption and leisure trends, as well as in the domain of life–work balance. For them, environmental issues are far less relevant than they were for the post-materialists. Measured in terms of standard questionnaires designed by environmental sociologists (with responses such as 'Yes, I would make sacrifices for the environment, even if others do not'), members of this group perform rather badly.

However, this does not necessarily mean that the fate of the environment is not important to them. It could also indicate that well-established interview scales are outdated

---

10 See www.sinus-milieus.de (accessed October 2009).
11 Social milieus can be read as a modernised version of social class, as they combine social status with values and everyday cultural practices.
12 These two terms mean different things. However, in social discourses, the scientific term 'ecology' has been transformed (during the 1970s and early 1980s) into a value-based political concept.
13 See www.sinus-institut.de.

and that we have to face lifestyle changes as part of a sustainability transition. Members of the 'modern performer' segment, for example (better-off professionals with liberal attitudes), expect societies and technologies to be environmentally friendly, and they would like to see governmental regulation and professional systems (including codes of conduct) in place in order to deal with the environmental side of societies' metabolism in a sustainable and professional manner.

One might interpret these wishes as an expression of the famous 'attitude–behaviour gap', or as another example for the not-in-my-backyard (NIMBY) syndrome. We would suggest another reading, however. To put it in a slightly exaggerated manner: modern performers display a rational attitude towards social complexity, the division of labour and the historical evolution of modernity. They believe that the individual consumer cannot solve problems concerning the collective good, that this instead requires public decision-making. In other words: their rejection of proactive individual engagement on environment problems reflects the difference between the consumer and the citizen—and correctly so. From a social science point of view, it would simply be naïve and overburdening if we were to expect individual consumers to behave in ways that were paradigmatically realised by the early 'alternatives', for whom ecology has been crucial in their personal and political life, if not an ideology. To articulate expectations about governmental action and to expect a professional approach to environmental problems is both a rational and an historically plausible attitude given the evolution of the issue over the past 30 years.

This is not meant to be an apologia for the 'modern performer' or other modern social groups. He or she displays consumption and leisure behaviours that 'consume' significant amounts of resources, and their behaviours indeed can be improved. Our point is a different one: the social diffusion of an issue (an idea, a technology or behaviour) across society is not congruent with the transfer of an identical 'thing' from one hand to another. The subjects and the issue change, especially if the issue is as complex and far reaching as 'the environment' (or, more accurately, our interactions with the environment). Today, in Germany, after the 'normalisation' of ecology, there is not only a 'green environmentalism' but also a conservative and a postmodern environmentalism.[14]

It is not simply the case that different social actors select different aspects of an issue; it is also true that conservatives select issues such as nature conservation or the need to conserve water, whereas post-materialists prefer renewable energy, and modern per-

---

14 For reasons of brevity we cannot illustrate this statement. The rise of organic food in many Western countries, especially in the UK, the USA and in Germany, would provide another example. The organic movement, with a history dating back to the beginning of the 20th century, experienced similar rigidities and was supported by similarly marginal groups as renewable energy is today. Over time, advocacy groups were able to influence national and international (European Union) legislation (regarding, for example, labelling, production standards and subsidies), changing the boundary conditions of the system. Growing consumer demand together with citizen-based policy reforms have led to intensified growth in this sector. Distribution channels have changed accordingly (from farmer markets to supermarkets) and price premiums have diminished as a result of economies of scale. However, the pressure on organic farmers is growing too, and by now many pioneers are asking themselves whether this systemic change was really intended—or if it would not be better to go back to the 'pure' origins of the movement.

formers ecodesign and health food. More importantly, all these groups adhere to differ-ent interpretations of what it means to be 'pro-environmental', and they frame the issue differently. Conflicts no longer arise between environmental and non-environmental attitudes (exclusively), but mainly between different interpretations of what environ-mentalism means. This has implications for design, communication, marketing and policy preferences. The green issue is changing its Gestalt, that is, its content, shape, interpretation and framing. Environmental social scientists should, in a way, be happy about this change, but it requires their special attention, as well-established instru-ments and approaches might unexpectedly become inappropriate for use.

We would like to make an additional point before we conclude. Most diffusion mod-els assume an homogeneous population. In this case, diffusion over time has the usual S-shaped form. If, however, the population is inhomogeneous, (for example, owing to different network densities), the overall diffusion rate will look different. In social real-ity, people differ in terms of their milieu affinity and lifestyles, and these distinctions (Bourdieu 1976) affect the overall speed and social localisation of diffusion.

This is an additional argument for more lifestyle-oriented research on environmental consumption. Non-linear systems behaviour might occur, and for sustainability policies it might be relevant to learn what diffusion pathways might occur and whether there are 'tipping points' (Gladwell 2000) for strategic decisions.

10.6 Conclusions

This chapter underlines the necessity for a sustainability transition. Business-as-usual strategies or even incremental (technological) changes will in no way suffice to reduce the detrimental environmental impacts of modern societies. Catch-up processes and globalisation aggravate the situation and stress the necessity of substantial changes. Many scholars (mostly mainstream economists) and politicians seem to assume that more or less radical changes in the technological sub-structure of modern societies will be adequate, with no substantial policy changes, no lifestyle changes. We argue against such an assumption. Even radical technological changes will occur only if poli-cies change and if these policies are supported by citizens and accepted by consumers.

We propose a need to abandon the narrow view of consumption as an individual pro-cess of purchase and use and instead to contextualise consumption, both with regard to the environment and with regard to the social reality of modern societies. The con-cept of lifestyle can be used to provide such a conceptual framework at the social meso level.

We suggest it is necessary to take the influence of consumers and citizens over tech-nologies and policies into account, as highlighted by case studies from the German wind energy sector, the diffusion of green lifestyles over time and recent initiatives for product carbon footprints. No individual case resembles the other, and so it would not be wise to draw general conclusions from the German case alone. However, a closer look at analogous cases in other countries might reveal similarities and differences—and that is important for mutual learning.

References

Anheier, H., M. Glasius and M. Kaldor (eds.) (2001) *Global Civil Society 2001* (Oxford, UK: Oxford University Press).

Bechberger, M., and D. Reiche (2005) 'Europe Banks on Fixed Tariffs', *New Energy* 2 (April 2005): 14-18.

—— and D. Reiche (eds.) (2006) *Ökologische Transformation der Energiewirtschaft Erfolgsbedingungen und Restriktionen* (Berlin: Erich Schmidt Verlag).

Beck, U. (1992) *Risk Society: Towards a New Modernity* (London: Sage).

—— (1996) 'World Risk Society as Cosmopolitan Society? Ecological Questions in a World of Manufactured Uncertainties', *Theory, Culture & Society* 13.4: 1-32.

——, A. Giddens and S. Lash (1999) *Reflexive Modernisation* (Cambridge, UK: Polity Press; Palo Alto, CA: Stanford University Press).

Bhalla, A.S., S. Yao and Z. Zhang (2003) 'Causes of Inequalities in China, 1952 to 1999', *Journal of International Development* 15: 939-55.

BMU (Bundesministerium für Umwelt, Naturschutz und Reaktorsicherheit) (2005a) 'Erneuerbare Energien in Zahlen: Nationale und internationale Entwicklung. Dezember 2005', internet update; www.erneuerbare-energien.de/files/pdfs/allgemein/application/pdf/erneuerbare_energien_zahlen_dezember.pdf (accessed October 2009).

Botzen, W.J.W., J.M. Gowdy and J.C.J.M. van den Bergh (2008) 'Cumulative CO_2 Emissions: Shifting International Responsibilities for Climate Debt', *Climate Policy* 8: 569-76.

Bourdieu, P. (1976) *Distinction: A Social Critique of the Judgement of Taste* (Cambridge, MA.: Harvard University Press).

Brand, K.-W., K. Eder and A. Poferl (1997) *Ökologische Kommunikation in Deutschland* (Opladen, Germany: Westdeutscher Verlag).

Buttel, F.H. (2003) 'Environmental Sociology and the Explanation of Environmental Reform', *Organization & Environment* 16.3: 306-44.

Byzio, A., H. Heine, R. Mautz and W. Rosenbaum (2002) 'Zwischen Solidarhandeln und Marktorientierung. Ökologische Innovation in selbstorganisierten Projekten: autofreies Wohnen, Car Sharing und Windenergienutzung', Soziologisches Forschungsinstitut der Georg-August-Universität, Göttingen, Germany.

Campbell, C. (1987) *The Romantic Ethic and the Spirit of Modern Consumerism* (Oxford, UK: Basil Blackwell).

Carrigan, M., I. Szmigin and J. Wright (2004) 'Shopping for a Better World? An Interpretative Study of the Potential for Ethical Consumption within the Older Market', *Journal of Consumer Marketing* 21.6: 401-17.

Christensen, P. (1997) 'Different Lifestyles and their Impact on the Environment', *Sustainable Development* 5: 30-35.

Cogoy, M. (1999) 'The Consumer as a Social and Environmental Actor', *Ecological Economics* 28: 385-98.

Cohen, M.J., A. Comrov and B. Hoffner (2005) 'The New Politics of Consumption: Promoting Sustainability in the American Marketplace', *Sustainability: Science, Practice and Policy* 1.1: 1-19.

Consumers International (1997) 'Discerning Middle Class? Sustainable Consumption: A Preliminary Enquiry of Sustainable Consumption Trends in Selected Countries in the Asia–Pacific Region' (Consumers International).

Cowan, R.S. (1983) *More Work for Mother: The Ironies of Household Technology from the Open Hearth to the Microwave* (New York: Basic Books).

Crocker, D., and T. Linden (eds.) (1998) *Ethics of Consumption: The Good Life, Justice, and Global Stewardship* (Lanham, MD: Rowman & Littlefield).

Diekmann, A., and P. Preisendörfer (2003) 'Green and Greenback: The Behavioural Effects of Environmental Attitudes in Low-cost and High-cost Situations', *Rationality and Society* 15.4: 441-72.

Doubleday, R. (2004) 'Institutionalising Non-governmental Organisation Dialogue at Unilever: Framing the Public as "Consumer-citizens" ', *Science and Public Policy* 31.2: 117-26.

Duchin, F. (1998) *Structural Economics: Measuring Change in Technology, Lifestyles, and the Environment* (Washington, DC: Island Press).

Durning, A. (1992) *How Much Is Enough? The Consumer Society and the Future of the Earth* (New York: W.W. Norton).

Environics (2002) *Consumerism: A Special Report* (Toronto: Environics International).

Fine, B., and E. Leopold (1993) *The World of Consumption* (London: Routledge).

Fischer-Kowalski, M., and C. Amann (eds.) (2001) 'Societal Metabolism and Human Population', *Population and Environment: A Journal of Interdisciplinary Studies* 23.1 (Special Issue).

Fuchs, D.A., and S. Lorek (2005) 'Sustainable Consumption Governance: A History of Promises and Failures', *Journal of Consumer Policy* 28: 261-88.

Giddens, A. (1986) *The Constitution of Society: Outline of the Theory of Structuration* (Berkeley, CA: University of California Press).

Gladwell, M. (2000) *The Tipping Point: How Little Things Can Make a Big Difference* (Boston: MA: Little, Brown).

Guinee, J.B. (ed.) (2002) *Handbook on Life-cycle Assessment* (Dordrecht, Netherlands: Kluwer Academic).

Hall, P.A., and D.W. Soskice (eds.) (2001) *Varieties of Capitalism: The Institutional Foundations of Comparative Advantage* (Oxford, UK: Oxford University Press).

Hernes, G. (1976) 'Structural Change in Social Processes', *American Journal of Sociology* 38.3: 513-46.

Hertwich, E., T. Briceno, P. Hofstetter and A. Inaba (eds.) (2005) 'Sustainable Consumption: The Contribution of Research' (Report 1/2005; Oslo: Industrial Ecology Program, Norwegian University of Science and Technology).

Huber, J. (2000) 'Towards Industrial Ecology: Sustainable Development as a Concept of Ecological Modernisation', in M. Andersen and I. Massa (eds), 'Special Issue: Ecological Modernisation', *Journal of Environmental Policy and Planning* 2: 269-85.

Hughes, T. P. (1983) *Networks of Power: Electrification in Western Society, 1880–1930* (Baltimore, MD: The John Hopkins University Press).

IPCC (Intergovernmental Panel on Climate Change) (2007) 'Climate Change 2007: Synthesis Report'; www.ipcc.ch/pdf/assessment-report/ar4/syr/ar4_syr.pdf (accessed October 2009).

Jackson, T. (2004) 'Models of Mammon: A Cross-disciplinary Survey in Pursuit of the "Sustainable Consumer" ', ESRC Sustainable Technologies Programme Working Paper 2004/1, Centre for Environmental Strategy, University of Surrey, Guildford, UK.

—— (2005) 'Motivating Sustainable Consumption: A Review of Evidence on Consumer Behaviour and Behavioural Change', a report to the Sustainable Development Research Network, Centre for Environmental Strategy, University of Surrey, Guildford, UK.

Jacobsson, S., and V. Lauber (2006) 'The Politics and Policy of Energy System Transformation: Explaining the German Diffusion of Renewable Energy Technology', *Energy Policy* 34: 256-76.

Kates, R.W., W.C. Clark, R. Corell, J.M. Hall, C.C. Jaeger, I. Lowe, J.J. McCarthy, H.J. Schellnhuber, B. Bolin, N.M. Dickson, S. Faucheux, G.C. Gallopin, A. Grübler, B. Huntley, J. Jäger, N.S. Jodha, R.E. Kasperson, A. Mabogunje, P. Matson, H. Mooney, B. Moore III, T. O'Riordan and U. Svedin (2001) 'Environment and Development: Sustainability', *Science* 292 (27 April 2001): 641-42.

Kemp, R., D. Loorbach, and J. Rotmans (2007) 'Transition Management as a Model for Managing Processes of Co-evolution Towards Sustainable Development', *International Journal of Sustainable Development and World Ecology* 14.1: 78-91.

——, J. Schot and R. Hoogma (1998) 'Regime Shifts to Sustainability through Processes of Niche Formation: The Approach of Strategic Niche Management', *Technology Analysis and Strategic Management* 10.2: 175-95.

Lange, H., and L. Meier (2009) *The New Middle Classes. Globalizing Lifestyles, Consumerism and Environmental Concern* (Berlin: Springer).

Lebel, L. (2005) 'Transitions to Sustainability in Production–Consumption Systems', *Journal of Industrial Ecology* 9.1-2: 11-13.

——, S. Lorek and R. Daniel (eds.) (2010) *Sustainable Production Consumption Systems: Knowledge, Engagement and Practice* (Dordrecht, Netherlands: Springer).

Lodziak, C. (2002) *The Myth of Consumerism* (London: Pluto Press).

Loorbach, D. (2010) 'Transition Management for Sustainable Development: A Prescriptive, Complexity-Based Governance Framework', *Governance* 23.1: 161-83.

Lüdtke, H. (1989) *Expressive Ungleichheit: Zur Soziologie der Lebensstile* (Opladen, Germany: Leske + Budrich).

Lundvall, B.-A. (ed.) (1995) *National Systems of Innovation: Towards a Theory of Innovation and Interactive Learning* (London: Frances Pinter).

Lutzenhiser, L. (1993) 'Social and Behavioural Aspects of Energy Use', *Annual Review of Energy and the Environment* 18: 247-89.

—— and B. Hackett (1993) 'Social Stratification and Environmental Degradation: Understanding Household CO_2 Production', *Social Problems* 40.1: 50-73.

Manoochehri, J. (2002) 'Post-Rio "Sustainable Consumption": Establishing Coherence and a Common Platform', *Development* 45.3: 47-53.

MGI (McKinsey Global Institute) (2006) *From 'Made in China' to 'Sold in China': The Rise of the Chinese Urban Consumer* (Los Angeles, CA: McKinsey).

—— (2007) *The 'Bird of Gold': The Rise of India's Consumer Market* (Los Angeles: McKinsey).

Müller, H.-P. (1992) *Sozialstruktur und Lebensstile: Der neuere theoretische Diskurs über soziale Ungleichheit* (Frankfurt am Main, Germany: Suhrkamp).

Myers, N., and J. Kent (2003) 'New Consumers: The Influence of the Affluence on the Environment', *Proceedings of the National Academy of the Sciences* 100.6: 4,963-68.

NRC (National Research Council) (1999) *Our Common Journey: A Transition toward Sustainability* (Washington, DC: National Academy Press).

OECD (Organisation for Economic Cooperation and Development) (2002) *Towards Sustainable Household Consumption? Trends and Policies in OECD Countries* (Paris: OECD).

Porter, M. (1998) *The Competitive Advantage of Nations* (New York: The Free Press).

Princen, T., M. Maniates and K. Conca (eds.) (2002) *Confronting Consumption* (Cambridge, MA: MIT Press).

Redclift, M. (1996) *Wasted: Counting the Costs of Global Consumption* (London: Earthscan Publications).

Reisch, L., and I. Ropke (2005) *The Ecological Economics of Consumption* (Cheltenham, UK: Edward Elgar).

Reusswig, F. (1994) *Lebensstile und Ökologie* (Frankfurt am Main, Germany: Verlag für Interkulturelle Kommunikation).

——, H. Lotze-Campen and K. Gerlinger (2005) 'Changing Global Lifestyle and Consumption Patterns: The Case of Energy and Food', in G. Radhakrishna (ed.), *Consumer Behaviour: Effective Measurement Tools* (Hyderabad, India: The ICFAI University Press): 197-210.

Robison, R., and D.S.G. Goodman (eds.) (1996) *The New Rich in Asia: Mobile Phones, McDonald's and Middle-class Revolution* (London: Routledge).

Røpke, I. (2004) 'The Early History of Modern Ecological Economics', *Ecological Economics* 50.3–4: 293-314.

Rotmans, J., R. Kemp and M. van Asselt (2001) 'More Evolution than Revolution: Transition Management in Public Policy', *Foresight: The Journal of Future Studies, Strategic Thinking and Policy* 3.1 (February 2001): 15-31; www.icis.unimaas.nl/publ/downs/01_12.pdf (accessed October 2009).

Schaefer, A., and A. Crane (2005) 'Addressing Sustainability and Consumption', *Journal of Macromarketing* 25.1: 76-92.

Schellnhuber, H.-J., and S. Rahmstorf (2006) *Der Klimawandel* (Munich: C.H. Beck).

——, W. Cramer, N. Nakicenovic, T. Wigley and G. Yohe (eds.) (2006) *Avoiding Dangerous Climate Change* (Cambridge, UK: Cambridge University Press).

Schor, J. (1998) *The Overspent American: Upscaling, Downshifting, and the New Consumer* (New York: Basic Books).

Shove, E. (2003) *Comfort, Cleanliness and Convenience: The Social Organisation of Normality* (Oxford, UK: Berg).

—— and A. Warde (2002) 'Inconspicuous Consumption: The Sociology of Consumption and the Environment', in R.E. Dunlap, F.H. Buttel, P. Dickens and A. Gijswijt (eds.), *Sociological Theory and the Environment* (Lanham, MD: Rowman & Littlefield): 230-51.

Spaargaren, G. (2003) 'Sustainable Consumption: A Theoretical and Environmental Policy Perspective', *Society and Natural Resources* 16.8: 687-701.

Sridharan, E. (2004) 'The Growth and Sectoral Composition of India's Middle Class: Its Impact on the Politics of Economic Liberalisation', *India Review* 3.4: 405-28.

Stern, P.C., T. Dietz, V.W. Ruttan, R.H. Socolow and J.L. Sweeney (eds.) (1997) *Environmentally Significant Consumption: Research Directions* (Washington, DC: National Academy Press).

Stevenson, N. (2002) 'Consumer Culture, Ecology and the Possibility of Cosmopolitan Citizenship', *Consumption, Markets and Culture* 5.4: 305-19.

Stier, B. (1999) *Staat und Strom: Die politische Steuerung des Elektrizitätssystems in Deutschland 1890–1950* (Heidelberg, Germany: Verlag für Regionalkultur).

The Carbon Trust (2007) *Carbon Footprints in the Supply Chain: The Next Step for Business* (London: The Carbon Trust).

Tomlinson, M. (2003) 'Lifestyle and Social Class', *European Sociological Review* 19.1: 97-111.

Uusitalo, L. (1986) *Environmental Impacts of Consumption Patterns* (Aldershot, UK: Gower).

Van Wessel, M. (2004) 'Talking about Consumption: How an Indian Middle Class Dissociates from Middle-Class Life', *Cultural Dynamics* 16.1: 93-114.

Walker, G., and D. King (2008) *The Hot Topic: What We Can Do About Global Warming* (Orlando, FL: Harvest Harcourt).

Weber, C., and A. Perrels (2000) 'Modelling Lifestyle Effects on Energy Demand and Related Emissions', *Energy Policy* 28: 549-66.

Weber, M. (1904) 'Die protestantische Ethik und der "Geist" des Kapitalismus', *Archiv für Sozialwissenschaft und Sozialpolitik* 20: 1–54 .

—— (1905) 'Die protestantische Ethik und der "Geist" des Kapitalismus', *Archiv für Sozialwissenschaft und Sozialpolitik* 21: 1–110.

Weingart, P., S. Engels and P. Pansegrau (2000) 'Risks of Communication: Discourses on Climate Change in Science, Politics, and the Mass Media', *Public Understanding of Science* 9: 261-83.

Wippermann, C. (2005) *Die soziokulturelle Karriere des Themas 'Ökologie': Eine kurze Historie vor dem Hintergrund der Sinus-Lebensweltforschung* (Heidelberg, Germany: Sinus Sociovision; www.sinus-sociovision.de/Download/karriere_oekologie.pdf [accessed October 2009]).

York, R., E.A. Rosa and T. Dietz (2004) 'The Ecological Footprint Intensity of National Economies', *Journal of Industrial Ecology* 8.4: 139-54.

Zukin, S., and J.S. Maguire (2004) 'Consumers and Consumption', *Annual Review of Sociology* 30: 173-97.

11
Conclusions: steps towards more sustainable energy use in housing

Saadi Lahlou
EDF R&D, Clamart, France; London School of Economics and Political Science, London UK

Tim Woolman and Martin Charter
The Centre for Sustainable Design, University for the Creative Arts, Farnham, UK

Arnold Tukker
TNO Built Environment and Geosciences, The Netherlands;
Norwegian University of Science and Technology (NTNU), Department of Product Design

In this concluding chapter we try to help stakeholders find a path for transition toward more sustainable energy use in housing, based on what we learned from the case studies in this book and more generally from the SCORE! Research programme. In this chapter we will review the problem then provide a new framework to help actors plan change.

The framework is a three-layered model of what determines individual behaviour: physical affordances (building and equipment), people's representations and practice, and the institutional rules of the game.

In Section 11.1 we review the nature of the problem. Rigidity and inertia are the main characteristics of the system, for various reasons which we summarise in Sections 11.1.1 and 11.1.2. Owing to this inertia, the reorientation of the system is slow. We list the major trends and also suggest some desirable trends that are weak or absent.

Following this review we take in Section 11.2 a grounded approach to understand the mechanisms of change in the domain. We compare change strategies in general with what we can actually observe in the field from the case studies of change. These fall into three different categories: local experiments, innovative communities and larger or non-local innovation. This is discussed in more detail through a review of common patterns observed in the case studies.

We can then consider the potential for change in the domain (Section 11.3). In stable sustainable systems, individual behaviour is regulated at three levels: physical infra-structure; representations and practice; and institutions—the so-called triple-determi-nation framework (Section 11.3.1). All three regulating layers must be addressed in any transition to a new system. The problem for the actors can be compared to 'greening' parts of that three-layered system sufficiently to operate and sustain the transition. In section 11.3.2 we provide the metaphor of a three-layered 'leopard skin' to describe how the global system can be 'greened', based on the case studies described in this book.

In Section 11.4 we offers a short comparison with the mobility and agriculture and food domains.

Notable issues and recommendations for action towards sustainable consumption and production (SCP) in energy use in houses and buildings are suggested in Section 11.5, including a list of critical points for success and failure in the typical three types of innovation: local experiment, innovative community, and larger or non-local innova-tion.

11.1 Systemic description of energy use in the housing domain

In this section we sketch the characteristics of housing and related energy use in con-sumer societies, the regime (socioeconomic organisation of the domain) and its impli-cations and current trends. We limit the scope of this analysis to 'consumer economies'; the problem being somewhat different in other economies (for a discussion, see Tukker *et al.* 2008: 4-5).

11.1.1 The landscape: relevant meta-factors and sustainability issues in the domain of energy use in housing

The 160 million buildings in the EU use over 40% of Europe's energy and create over 40% of its carbon dioxide (CO_2) emissions, and that proportion is increasing, as aver-age energy consumption grows by around 0.4% per year

Chapter 2 described the major 'meta-factors' influencing the landscape of the domain, notably societal evolution such as ageing and individualism. As a brief reminder, the population of Europe is increasing, as is the number of households, partly as a result of an ageing population. More numerous and smaller households. each affording and using more energy-using products, are driving increases in overall energy use per capita, despite improvements in the energy efficiency of appliances. 'Informatisation' is slowly bringing home automation and with it some control on demand, responding to rising prices and also to the desire for comfort. The demand for maintaining thermal comfort in buildings, for both heating and cooling, is also affected by local climatic change. Higher levels of demand, with greater degrees of variation, are already putting strain on the energy generation and distribution system. Price fluctuations, mostly price rises, and media speculation are sensitising political debate of energy and related policies. Political authorities are still broadly reluctant to adopt unpopular measures affecting the consumption and production of energy, despite wider recognition and acceptance of the need to make changes towards sustainability.

The gap between declared awareness of sustainability issues and actual behaviour persists—although this is lessening as media interest is beginning to focus on the behavioural aspects of sustainability and energy price rises are becoming pervasive; also, these factors are having knock-on effects to other domains such as agriculture and food and mobility.

People remain inherently reluctant to change their habits, especially in something as private as the home. More sustainable consumer behaviour could come through energy-saving habits (for example, by choosing appropriate thermostat settings, switching off lights, etc.) or buying more efficient appliances (low-energy light bulbs, energy-efficient washing machines, etc.), but changes in niches are slow to spread to the mainstream. Resource pooling in product–service systems has also had limited penetration.

Working with these meta-factors and convincing and supporting the present stock of stakeholders in the regime, notably inhabitants, to reduce energy-intensive consumption, seems necessary given the need to make fundamental changes in a fraction of a generation.

The next section will re-examine the regime of energy use in housing domain and how it is organised. This section summarises Chapter 2, revisiting the regime in light of the case studies presented in Chapters 3–10.

11.1.2 Energy use in housing and interlinked practices: the 'regime'

In re-examining the socio-technical regime of energy use in houses and buildings, inertia is a strong theme. This is a consequence of the rigidity of the infrastructure, of the ongoing demand for comfort within buildings and the connectedness of these with many other sectors of activity and of the large number and high price of buildings. Indeed, as noted in Chapter 2, the price of buildings in the regime, which is based more on the price of land than of construction itself, adds to the inertia by making any decision laden with its economic impact.

11.1.2.1 Aspects of the regime

Consumer economies have a large stock of existing buildings; energy supply infrastructure and institutions are mature; people are concentrated in dense and historically structured urban areas.

Most existing buildings are not energy-efficient, and it is difficult and expensive to retrofit an old building to make it energy-efficient. Constructing a new house takes between 3 and 5 years (including study and decision time); a complex retrofit may take 10 years. Each case being specific, there is little economy of scale in retrofitting buildings. Although it is more efficient to build a new building rather than retrofit an old building, this option is rarely considered. Buildings are designed to be operational on a continuous basis as a life and activity support system, with the associated energy demand largely 'built-in'. They cannot easily be disassembled or elements replaced. Change opportunities for buildings are usually limited to specific 'windows of opportunity' when people move or change the configuration of their household, often coinciding with life changes or when the building envelope or infrastructure needs a major update.

Current infrastructure is often an obstacle to more sustainable energy provisioning for buildings. Most construction is urban, high-density housing, which can often exclude or limit some uses of renewable energy (such as wind power). The relation of local renewable energy to the energy grid (sometimes suppliers, sometimes consumers) may need a deep restructuring of the electricity grid, which is constrained by vested interests.

When buildings are designed and constructed, energy efficiency is only one concern among many—the focus is on construction costs, not on running costs. However, the recent significant rises in fuel prices, as a result of rising oil prices, are stimulating a wider awareness of energy use in houses and buildings and alternative energy (cost) saving measures.

Given most modern human activity takes place indoors, the societal function and physical nature of these buildings as they are currently (culturally) constructed also accounts for many of the difficulties encountered in promoting change, both in the present and in the future. Through this meshing of the building structure with its internal infrastructure, most changes in the building itself may have implications on the behaviour and practices within the building, which may spread to other domains. Therefore, modifications are difficult on a social as well as a technical level. Changing something in a house feels like trying to mend the hull of a boat at sea.

Public authorities have provided more incentives and regulation for sustainability, for example through the EU Emission Trading Scheme, and have deregulated the energy market (see the discussion of market institutions below).

Examining the norms associated with regulatory institutions, such as the Energy Performance of Buildings Directive (EPBD; European Council 2003), shows the complexity of issues in the system for regulating the rating and certification of buildings to meet energy performance requirements (illustrated in Chapter 2). In addition to the EPBD, the Energy Services Directive requires member states to adopt a target of a 9% saving in energy end-use by 2016 and to put in place the institutional and legal frameworks and measures needed to remove barriers to efficient energy end-use:

> It is intended to act as a catalyst for renewed and more ambitious energy efficiency initiatives at all levels of European society—local, regional, national and Community. It should create the necessary conditions for the development and promotion of a market for energy services and the delivery of energy efficiency to end-users (European Council 2006).

The market is the main mode of 'regulation' in this domain, which is both a favourable ground for local innovation and an obstacle to long-term public policies. The electricity distribution infrastructure and market is still a crucial element in the system. An 'intelligent grid', enabling distributed, local and renewable power generation will be a major component of increased efficiency and sustainability but is still in the making (Harris 2008). Hence, in European consumer economies, a major characteristic of the housing regime is rigidity. Buildings are fixed assets; they are durable, they are not technically designed for easy evolution or retrofit. The housing system is considered as a fixed framework sheltering various activities, expected to provide ambiance control. This has a deep influence on most aspects of our sedentary life.

This rigidity is both an obstacle to change and, conversely, a guarantee that what is changed will have a pervasive ongoing impact across society. So, in approaching the challenge, it is especially crucial to recognise the sustainability problems and exploit the windows of opportunity for change.

11.1.2.2 Exploiting windows of opportunity: current trends in the domain

As discussed in Chapter 2, windows of opportunity, together with some risks, come from trends in the context and landscape and regime, forcing producers and consumers in a specific direction. Rising oil prices, consumer demand, public policies and regulation, global warming and media hype create the solid anticipation of a market for sustainable solutions, which is a positive driver for innovation and investment. Deregulation of the energy market and technical progress are not only driving but also destabilising factors. They enable local innovation but make long-term decisions more difficult because they multiply the number of players with a short-term horizon and thus add uncertainty.

Some trends of change are visible and some that might have been expected are absent or barely discernible. The first positive and visible trend is the development of 'zero-energy' houses in new construction. The second is greater energy efficiency (people are consuming less) in existing housing, with new energy-use practices. The third is the emergence of an 'intelligent' energy grid incorporating distributed and renewable generation. The first two areas of change are conducted by actors in the building and housing sector whereas the third is conducted mostly by energy service companies (ESCOs) and energy companies.

11.1.2.2.1 'Zero-energy' (new) eco-houses and buildings, and eco-renovation

Zero-energy buildings aim to reduce the consumption of non-renewable energy to less than the amount they produce during use. There are many variations in the strategies and technologies used, which accounts for the many descriptions applied to zero-energy buildings, such as 'passive houses', 'building as power plant', 'zero-energy developments', 'carbon neutral buildings' and so on. Techniques to achieve this end include: use of insulation; careful monitoring of thermal management (through double-glazed windows, green roofs, shading, electronic regulation with distributed sensors, etc.); use of locally generated power or heat (through wind turbines, photovoltaic cells, passive solar energy, etc.); wise use of thermodynamics (through thermal mass, energy cascades, heat pumps, etc.); and use of locally affordable resources in an energy pool with neighbouring buildings. Through this growing market the technologies applied to new and existing buildings are now experiencing accelerated development. However, they remain easier to install in new construction. Many of the elements brought together in zero-energy houses are still at a niche stage, but the market for them, including as combined technologies, is beginning to increase in some regimes, especially where there is public financial help available or regulatory incentives; for example, steps are being taken towards making new homes in the UK 'zero carbon in use' by 2016. By making energy performance diagnosis and certification compulsory, the EPBD directly influences the price of a building and is therefore a strong incentive for sellers or owners to upgrade their energy efficiency.

11.1.2.2.2 Better energy efficiency and demand-side management

Better energy efficiency and demand-side management is achieved by modifying consumers' behaviour and their adoption of efficient technologies in existing houses. Heating and electrical equipment in buildings are responsible for a large proportion of energy consumption in those buildings. Now, equipment is being made more efficient. The 1992 European Council Directive on the Indication by Labelling and Standard Product Information of the Consumption of Energy and Other Resources by Household Appliances (European Council 1992) set the framework for measures on refrigerators and freezers, washing machines, dishwashers and ovens. The aim is to use compulsory labels on products to promote to consumers the benefits of buying appliances that are more energy-efficient and helping consumers to calculate the return on investment of a product. Labels are to provide standard information on energy consumption of the appliance concerned. It is therefore also an incentive to manufacturers to go beyond the minimum standards and is intended to focus competition on environmental characteristics.

As well as the energy efficiency of appliances, user behaviour has a major impact on consumption and therefore has been the target of many policies. Demand-side management (DSM) is an old idea. DSM can be achieved, for example, by modulating electricity tariffs according to time of day. In 1965 in France, the energy company EDF started the first DSM tariff, called 'heures creuses', where electricity was cheaper during the night, which enabled consumers to heat their water at night when demand was lower. This was followed by another tariff modulated specifically to discourage users from consuming during 22 peak days per year. Such tariffs were created in order to avoid constructing more power plants to manage peak loads yet still meet consumer demand. DSM can also be achieved by providing meter feedback to consumers (see Chapter 9). In Europe, some energy providers are starting to offer intelligent meters, which will to some extent enable such consumer feedback. Home automation is a key issue here. It may bring huge progress in consumer feedback, and aid energy-use control, but most actors in the domain are still hesitating to use such strategies because the technology is not fully mature and there are, as yet, no 'killer services'.

11.1.2.2.3 Intelligent grid and green energy provision

Intelligent grid and green energy provision is aimed at creating a more flexible grid that can accept an increased rate of penetration of renewable energy sources, increase the participation of customers in their own energy choices and meet the more reliable and resilient energy supply required by the society (ETP 2009). The main ideas are as follows:

- Smart energy coordinated management: this develops the concept of a 'virtual power plant' to transform the weakness of dispersed resources (in terms of meeting production to demand) into a strength through a smart aggregated offering of commercial and technical services (cf. the European Fenix project)[1]

- Smart metering: smart metering is the first 'smart' technology to be implemented within homes with the objective of optimising energy consumption. Remote and more frequent metering is only one aspect of what this technology can enable: it also has the potential to offer load management, increased information to customers about their electricity use and the creation of a single communication infrastructure that could offer a backbone to attach other 'smart grid functions' (such as fault location and better asset monitoring)

- Other enabling ICT: the power industry has been very conservative in adopting new communications technology, but new widely used IT solutions are available to modernise the grid (offering communication infrastructure, radio frequency identification [RFID], mobility solutions, the potential for detailed data-stream analysis or mining so as to have a grid that will be able to manage both distributed generation and home-level demand-driven optimisation, and so on)

This area needs not only more research but also to demonstrate the value of the deployment of such solutions and to encourage government to alleviate the regulatory barriers; as a paradox, in the EU the reorganisation of the electricity industry to open the market and create a fair and transparent market structure created rules that make more complicated the deployment of functions that would benefit society (such as load management). Also, these developments are in places hindered by the present trend of deregulation because they entail significant long term investment, which were usually incurred by large state-owned utilities. The cases in Woking (see Chapter 6) and in some German towns show the feasibility but also the limitations of grid management that includes some local energy generation by ESCOs.

These trends are particularly relevant to the installation of different types of infrastructure, which can bring more efficiency without changing the everyday behaviour of end-users. A problem is that some potential trends are absent or weak. On the production side there is tension and uncertainty about the pace of progress towards the use of renewable energy—a weak trend progressing slower than expected, partly from misunderstandings about the possible pace of replacement of present fossil-fuelled energy generation. See Chapter 2 for a discussion about the issues of grid management and problems associated with distributed generation.

On the consumption side, there is an action–awareness gap, where actors are aware of and supportive of ambitions towards sustainability (values) yet generally fail to translate that into positive choices affecting consumption (practice).

---

1 For more information on the Fenix Project, go to www.fenix-project.org (accessed October 2009).

11.1.2.2.4 Product–service systems and shared services,

Product–service systems (PSSs) and shared services, both reducing resource intensity, do not seem to be emerging in this domain—in contrast to what happens to some extent in the areas of food and mobility. PSSs are an obvious way to achieve better efficiency. A PSS company that sells a service for a given price will have a strong economic incentive to minimise the consumption of energy, hardware and consumables associated with providing that service and to invest in more efficient and durable equipment, thereby minimising the ecological footprint. Cultural reasons, such as expectations for individual ownership, inhibit the development of this business model. This goes alongside the private nature of housing in our societies. For example, the shared use washing machines, although evident in Sweden, does not seem to be an idea that has disseminated to Europe. In general, the tendency to individualisation is an obstacle to collective practices that may save energy, space and equipment (such as shared freezers, dryers, kitchens, etc.). It also needs infrastructure, which is rarely planned for in new construction.

A major challenge is therefore to change a system that is very 'inert' and that has a long life-cycle; that is, it is connected to most of our daily routines and requires us to change it 'afloat'; a situation somewhat different, for example, from that in the food or mobility areas.

However, the fact that houses are fixed enables us to consider local geographic policies more easily because a local change will have little impact elsewhere in space. Also, as most stakeholders involved are engaged in the same geographical area it is easier to work out local compromises over the set period. The building domain is far from globalised.

Finally, the long life-cycle of buildings is an advantage when one considers the durability of solutions. It is difficult to deconstruct what has been done; so, in contrast to the situation in other domains, if a good solution is built then it will not easily be destroyed by a transient change in local conditions, by policy, by the failure of some stakeholder, by a change in regulation or by the performance in some distant part of the supply chain and so on.

Considering these factors and the current trends, the small turnover of 'legacy buildings' replaced by new construction, the minor share of new construction and the slow pace of upgrading the present housing stock, we cannot expect 'natural evolution' to provide sustainable solutions in the short or medium term; therefore, change must be actively fostered by actors. How?

11.2 Potential for further change towards more sustainable energy use in houses and buildings

Before examining the potential for change and how actors can foster that transition, let us take a grounded approach to see what forms the changes actually take in the domain and compare them with theories.

There is extensive literature on change policies and innovation in general. Let us focus here on two dimensions of change: the change strategies, and the levels of change.

11.2.1 Change strategies

Tukker *et al.* (2008) distinguish two different strategies: 'innovation systems' that foster innovation, and 'system innovation' that tries to orient the system by regulation. The problem with innovation systems is that they help in innovating but do not necessarily orient the innovation in a given direction which, in a market economy, may not always lead to sustainable solutions.

In the system innovation approach, innovation is not a goal in itself but a means towards some specific change. It is the system innovation approach that we will consider here, keeping in mind that 'innovation systems' may still be needed in that process, for example to foster the emergence of new models or actors to be channelled toward a more sustainable plan.

Change management in current system innovation policies tends to follow mostly the 'transition management' model:

> [Transition management] uses approaches such as visioning, learning by doing, doing by learning, and adaptive management by groups of front runners to increase the probability of change to sustainable systems. It does so by supporting the development of promising niches, taking measures that create pressure on the regime, and providing a sense of direction for front-runner actors (Tukker *et al.* 2008: 417).

11.2.2 Levels of change

Tukker *et al.* (2008) distinguish three levels in describing change: the micro-level ('niches'), the meso-level ('regimes') and the macro-level ('landscape'), described as follows (Tukker *et al.* 2008: 416):

- Micro-level (niches): these are radical novelties that are not yet widespread but survive in protected spaces, such as small markets, where very specific values are relevant

- Meso-level (regimes): these are a set of interdependent and co-evolving technologies, symbolic meanings, infrastructures, consumer practices, institutions and expectations that reflect the mainstream way of doing things in a specific field

- Macro-level (landscape): this consists of very stable boundary conditions that cannot or that only with difficulty can be influenced by the regime, for example geopolitical realities such as the location of oil resources

The question of how to foster change seems to present itself differently at each of these different levels of change. For example, at the niche level one can assume the inside of the niche can be influenced whereas the outside environment can be considered as an intangible constraint. At the macro-level, change seems almost impossible from an individual actor's perspective; it needs the enrolment of many stakeholders. The situation at the meso-level is in between, but the variables are many and complex to monitor.

Now let us see what kind of actual cases we find in the domain of energy use in housing. The next section provides a partial review and classification of the various steps of transition that we have observed in the cases compiled through the SCORE! project in the domain of energy use in houses and buildings. Most of the cases we mention here have been described in detail in Chapters 3–10; for the others, we try to provide enough details of key elements.

11.2.3 What we see in the SCORE! cases: three typical situations

The cases lead us to distinguish three different main situations. We can represent these in an idealised model of three steps for transition:

- Step 1: local experiments. These are small, isolated, points of change, where a sustainable building is constructed or is created by retrofitting an old building

- Step 3: innovative communities. These are larger but still localised initiatives (limited to a single town or city, say) and typically combine construction and energy generation

- Step 3: regime change. This step represents an attempt to change the regime at a larger and non-local scale (by companies, states, professional bodies and so on)

These situations do not map exactly onto the niche–regime–landscape framework because of the nature of the housing domain. As we have seen, the rigidity and inertia of the housing domain somewhat limits the impact of other socioeconomic trends in the landscape, and the fixed and durable nature of assets, as well as their connection with urban infrastructure, provide a strong geographical flavour to the levels of market or regulatory institutions.

11.2.3.1 Local experiments

The first typical 'niche' situation is a step from vision to local experiment. 'Champions', motivated entrepreneurs, often architects, driven by a vision, act as local innovators to drive local projects. At this stage, motivations are individualised. This is illustrated by the cases presented by Brown and Vergragt (a retrofit in Boston; Chapter 3), Wimmer and Kang (a straw-bale house in Austria; Chapter 4) and many other cases, such as the Hockerton Housing Project.[2] The case presented by Wüstenhagen (solar houses in Freiburg; Chapter 5) and that of BedZED (Beddington Zero Energy Development, a

---

2 For more information on the Hockerton Housing Project, go to www.hockertonhousingproject.org.uk (accessed September 2010).

zero-fossil-energy housing development in outer London)[3] are an extension of these local experiments, illustrating the integration of such larger experiments (several buildings) into the wider framework of a city. These local points of change become more and more frequent as the trends and windows of opportunity, discussed above, combine with factors favouring potential change and actions by actor groups, suggested in the following sections.

Chapter 3, by Brown and Vergragt, and also the BedZED and Hockerton Housing Project cases presented at the SCORE! workshop in Paris (Lahlou and Emmert 2007), illustrate the crucial importance of the quality of vision, project leaders and teams in the success of innovation at the first level of local experimentation. For the present state of the art, the act of making a sustainable building is still an innovative project. Of course, many building projects are somehow innovative, but sustainability involves a need to change some classic routines, and this seems to cause a huge overhead in negotiation with the usual stakeholders, even though the technologies used may be not so complicated per se, and though they may have already been thoroughly tested in other places. The project team must spend extra energy, first to assemble off-the-shelf technology that has never been assembled before in that place (which brings extra cost and risk), then to convince all other stakeholders to share the risk that they think they are taking. This is why many sustainable buildings projects are often one-of-a-kind local experiments—bounded sociotechnical experiments, as Brown and Vergragt call them.

In this situation, the benefits of the innovative project must be considerable to justify 'buy-in' by stakeholders. Precisely, the benefits related to sustainability are of a kind that are rarely considered valuable; some might call them ideological.[4] Costs may not be considerably lower, and there may also be more constraints and sometimes less comfort, or more effort may be required.

Over an extended period, the actual financial or usability return on investment may be positive. For the BedZED project, for example, a full balance sheet was made available to stakeholders to counter misconceptions about potential return on investment.[5] The costs of setting up and maintaining an innovative project also remain visible, but the benefits of sustainability may need promotion to maintain ongoing exposure.

This means that the stakeholders who will be recruited at the present state of the art are those who either personally value the ideological benefits or are in some way interested in being involved in the experiments—for which there can be many reasons. As this step, the actors' behaviour cannot be explained by economic rationality alone. Just as in the case of the economics of culture and art described by (Lahlou *et al.* 1991), the economic actors are in fact driven by passionate personal motives that make them act beyond economic rationality. Here, this rationale may be a green ideology—hence the importance of individuals at this step.

---

3 For more information on BedZED, go to www.bioregional.com/what-we-do/our-work/bedzed (accessed September 2010).

4 This 'necessity' for political idealism, so indispensable during the pioneering phase, was also observed by Reusswig *et al.* (Chapter 10) in the beginning of the wind energy user groups in Germany in the 1980s. Then, governance switched from networks to economic markets.

5 Personal communication with Jess Hodge, Manager, One Planet Living in Sutton, BioRegional Development Group, 26 May 2007.

One may expect that cases increase, building for sustainability will spread towards becoming the norm, and these biases will fade away. One key issue is then to hasten the use of such good practice by capturing and making available successful experiences, not only of the final specifications of sustainable buildings but also the processes involved in sustainable building projects, especially the aspect of assembling the various technologies and expertise and assessing the balance of value for all project elements in a given context—an example is the application of Building Investment Decision Support (BIDS™), described by Loftness *et al.* in Chapter 8.

Apart from establishing the initial return on investment, the experience from the BedZED case shows the importance of building in ongoing maintenance funds to cater for the risks taken with new technology, in that case to enable the combined heat and power (CHP) unit to be renewed.[6]

Wimmer and Kang, in Chapter 4, trace the history of a project creating a sustainable house made from straw bales. This house achieves a Factor 10 improvement in resource efficiency and uses only local materials: straw, clay and wood. Beyond the technical interest of the house itself, Wimmer and Kang show how this house is used for demonstration and 'evangelising'. The 'S-House' has become a seminar centre where the straw-bale building technique is taught. Since the case is presented by the champions themselves, this case modestly does not emphasise the difficulties encountered, but the amount of persuasion and persistence that were necessary can be read between the lines. It is noteworthy that in some cases the champion had to solve issues far from his initial field of expertise, for example designing a specific screw for straw bales. In this case, as in others, it appears that the champion must take responsibility for a series of tasks that are, in classic building construction, usually distributed across various professionals in a conventional division of labour. This need for the creation of a new format for doing the work, and inventing solutions to 'fill in the gaps', is typical of these innovation situations. Innovators must create what Eymard-Duvernay and Thevenot (1983) called *investissements de forme*, 'investments in pattern'.

By 'investments in pattern' they mean the creation of some intermediate pattern that helps the actors to handle a complex situation. This intermediate pattern can be, for example, a cognitive category (a classification, a format), an institution (rules, conventions) or even technical tools. In other words, some of the cognitive work that has been done in solving problems can be capitalised and re-used in similar situations and therefore can be considered as an investment.

In Chapter 5, Rolf Wüstenhagen describes how a series of solar houses in Freiburg was built. He also shows the role of a 'champion', Rolf Disch, in the creation of this new generation of houses, which produce more energy than they consume. The story of how Disch implemented his designs is a story of a series of uphill negotiations to find financers; his initial project was sized down but finally reached commercial and media success. It was possible only with the support of the city of Freiburg and of visionary and partly philanthropic funders. In the meantime, Disch had to become a social entrepreneur and invent solar real estate funds. This case shows that the innovative project faced significant difficulties to receive funding in the classical real estate market, even

---

6 See footnote 5.

though the houses cost initially only 15% more than classic housing yet proved efficient by standards still 10 years after their construction. Many aspects of the project have now been taken on by prefabricated house manufacturer WeberHaus, which is a step into the niche market. Note that the state subsidy for solar energy in Germany also helped the project.

11.2.3.2 Innovative communities

The second situation is a step from local experiments to niche markets. Small communities of users and professionals, sharing the same ideology, strive to create local cultures that eventually become niche markets. In this situation, the role of the motivated individual appears crucial again, but this time they are mostly 'aediles' who drive a community rather than an individual project (local politicians, civil servants in regulatory institutions). The experience gathered produces a milieu of professional know-how, structured discourse and social capital. At this level, the motivations are to add social value in local communities and to raise status or gain recognition. Products and systems reach some level of usability but need strong commitment from participants. This social added value compensates for the shortcomings of the system for users. Regulation and governance systems for consumption patterns emerge; motivated professionals set up tools and good practice by trial and error. Public support through research grants and specific legislation enable business models that otherwise would not be sustainable.

This second situation is illustrated by the Hockerton Housing Project, by Loftness *et al.*'s BIDS™ system (Chapter 8) and by the case of Woking, presented by Thorp (Chapter 6); it is also demonstrated by a small but growing number of eco-towns and eco-cities, such as Dongtan in China (WBCSD 2006).

The Hockerton Housing Project is a five-house rural development in the UK and illustrates the case of a small community. It uses entirely passive solar heating, with a large thermal mass to stabilise temperatures, achieving a Factor 5 reduction in energy consumption for a build cost similar to that of conventional housing. Although it is smaller in size than the solar houses project described by Wimmer and Kang (Chapter 4), it is a multi-decision-maker project as the owners of the houses are independent. The five households that initiated the Hockerton Housing Project, without external funding, also share their experiences as a valuable educational resource, seeking to overcome the lack of belief that zero external fuel heating is possible. The evidence of the Hockerton Housing Project suggests that equivalent low-energy design principles could and should be incorporated into commercial developments and that similar projects would be viable without a particularly committed or cohesive group. The drivers behind the Hockerton Housing Project are trying to set up a network to get housing associations and builders involved in disseminating low-energy design.

Recognition of the achievement of the Hockerton Housing Project fostered local acceptance from the public and local authorities for renewable energy, speeding subsequent developments on the Project site—a second wind turbine was granted permission in a fraction of the time taken for the first—the process for the first turbine took four rounds of application before success.

In Chapter 6 Thorp describes how a local authority, Woking Borough Council, in the

UK, strives to reduce CO_2 emissions through large novel fuel cell and photovoltaic (PV) installations and private-wire local distribution. Woking Borough Council addresses the issue holistically through eight areas of concern: (1)planning and regulation, (2) energy services, (3) waste, (4) transport, (5) procurement, (6) education and promotion, (7) management of natural habitats and (8) adapting to climate change.

The Council set up a governance system for climate change initiatives, with decision and financial procedures, and, in 1999, established a group of companies (Thameswey Ltd) to contribute to the strategy in the market regime. This company has financial objectives, including a return on investment for shareholders (+8%). According to Thorp, the Council was successful as a result of a combination of factors, such as technical, financial and commercial innovation, partnering with the private and the use of a local electricity balancing and trading system.

The Council addresses various stakeholders: the fuel poor and the social workers who care for them as well as the planners and developers. For example, for the fuel poor, energy spending is maintained below 10% of state pension income; this is achieved in cooperation with the EnergyCare Network, which trains local social workers.

Thorp identifies the great number of institutional bodies that had to be involved in order to reach the goal. Stakeholders are usually grouped in communities of interest (for example, associations), and the regulation of their economic or social behaviour involves various technical or political bodies that must be involved. A quick count shows that formal negotiations with at least 20 different organisations were involved, along with applications to several government funds and programmes.[7] This means that the Council had to invest significant effort in engaging administratively with these various organisations in order to share visions and produce consensus.

Conducive (local or national) planning policies would have helped to develop a consensus around accepting the development in both the Hockerton Housing Project and the BedZED cases. Once BedZED was established as a successful demonstrator of a zero (fossil) energy development, principal partner, BioRegional, was consulted on national sustainable housing policy, and the BedZED case is used as an example to show that zero (fossil) energy housing is possible in a mainstream urban context. The principle followed in the design of BedZED by Bill Dunster Architects was to reduce energy demand as far as possible before looking to low-impact sources of supply. Zero (fossil) energy development (ZED) principles can be applied to mixed developments and retrofitting as well as to brownfield housing developments. An important part of the replication of ZEDs is establishing the (preferably local) supply chain. BioRegional have followed the BedZED experience by establishing One Planet Products to prepare suppliers not only for their own subsequent projects but also for others' projects and by

---

7 Among those involved were the Energy Saving Trust (EST), the Low Carbon Buildings Programme, Woking Borough Council, Thameswey Ltd, Xergi Ltd, the Lightbox Museum, Woking Theatre, Woking Borough Homes Ltd, the Community Energy Programme, the Department for Environment, Food and Rural Affairs (Defra), the English Partnership Commission for New Towns, Crest Nicholson and Frontier, Abbeygate and Sinsbury, Energy and Conservation Solar Centre Consulting and Beacon Council.

spreading the ZED principles to 'One Planet Living' buildings, forming partnerships in China, North America, Australia and South Africa.[8]

In this stage of extension to niche markets, public financial support helps again. The renewable obligations credit is a key aspect of the financial balance of Woking's wind power project (three large turbines), as for the UK government's other initiatives on micro-generation. Otherwise national developers, English Partnerships, covered a financial shortfall for the CHP development in Woking. Here we see how local actors take profit from state support, and conversely how a state funding support policy can orient and bias local experiments. This shows the nature of the institutional work innovators have to do to push transition from niche experiments to regime change and then beyond to mainstream markets. It takes a lot of effort to make the new niches compatible and competitive with the landscape of the present regime.

11.2.3.3 Larger or non-local changes

The third situation is less clear-cut. We could characterise it as a transition from niche markets to regime change. Based on the success of niche markets created by innovative communities, helped by mass media (champions are often good opinion leaders and also act as 'evangelists'), new solutions reach some threshold of social, political and economic visibility. The cluster of niche markets eventually becomes a large enough potential market to raise interest from larger stakeholders in the political and economic domain. A virtuous circle starts that produces a larger shift to the construction of a sustainable market. At this stage, economic or political motivation takes over and commercial companies are key actors. Products and services are easy enough to use to bring functional utility to users.

This situation is illustrated by the cases described by Loftness *et al.* (Chapter 8, on BIDS™), by Fischer (Chapter 9, on consumer feedback on electricity consumption) and by Reusswig *et al.* (Chapter 10, on wind energy in Germany). Together, these chapters show the complex ecology of stakeholders and technology that must be modified at this stage of broader diffusion. Most normative efforts are at the national or international level, such as building codes (such as those administered through the International Organisation for Standardisation [ISO],[9] the International Code Council [ICC 2004] and the American Society of Heating, Refrigerating and Air-conditioning Engineers/Illuminating Engineering Society of North America [ASHRAE/IESNA 2004]), EU Directives (such as the EPBD and EcoDesign of Energy-related Products [European Council 2003, 2009]) and national regulations. However, there have also been attempts to create a larger 'ecosystem', such as large companies aiming to produce global product lines, to introduce 'green' labels for energy and steps towards 'green' provisioning. As the ecologies considered here involve a large number of actors who may not necessarily be directly in touch with each other the nature of the system is complex, and actors have less knowledge and control.

In Chapter 8 Loftness *et al.* address the renovation market issue. The tool developed

---

8 See footnote 5.
9 www.iso.org

at Carnegie Mellon University, BIDS™, is an evidence-based online resource that enables the calculation of the return on investment of energy-saving architectural options. The decision to retrofit buildings is in the hands of investors and professionals. As in most professions where many regulations have historically been constructed and are based on good practice and state of the art, conservative behaviour is encouraged, because they can build on precedents to convince stakeholders that they are compliant with stakeholder interests. More sustainable buildings usually cost a little more in terms of initial financial investment, but they pay off in a few years because they cost less in energy consumption during use; also better buildings have positive effects on productivity (for example, through improved health and well-being). Costs are easy to add up, but benefits lie in the future and need to be evaluated. However, whereas new solutions need to prove their interest to financers, most of them do not have the proper decision tools in order to do so. Therefore new solutions that would be more sustainable prove difficult to evaluate in advance. They appear to contain more uncertainty or risk. Present evaluation by decision-makers typically is not based on sustainability but on economic return on investment from the perspective of the financers. The interest of the BIDS™ database and decision system is to address this block in the sustainable decision chain by feeding decision-makers with life-cycle performance based on actual cases, in an economic language that is used by those decision-makers.

In a typical application of BIDS™ the logic of reconstructing an artificial environment inside a building (artificial lighting, heating, ventilation, air conditioning, etc.), which normally induces high costs, can be partly avoided if there is recognition of the value of access to the natural environment, especially for natural lighting and ventilation. Loftness *et al.* show that BIDS™ can demonstrate over 30% reduction in lighting expenses, and 45–70% in cooling energy. This is also preferred by users and provides better comfort and health.

One issue with attempting to change behaviour is the need to establish a feedback loop so that consumers become aware of the consequences of their actions and can also compare their own practice with average or best practice. At a wider level, engaging a community of actors who can provide mutual support and recognition and providing evidence to understand the full consequences of change is also a demonstrably desirable step to promote the transition from local experiments and niches, where participants are driven largely by social or ideological motives, to the mainstream, where participants are driven predominantly by functional and economic motives.

Fischer (Chapter 9) reviews the influence of consumer feedback to stimulate electricity conservation. Fischer conducted an analysis of 26 case studies (half of them research projects) and makes technical recommendations on how to build up tools for consumer feedback. In this perspective, this work is similar to what BIDS™ does: providing stakeholders with decision tools that enable them to examine their behaviour based on evidence from actual cases. Fischer clearly shows that detailed metering and consumer feedback is a necessary step towards sustainability—it is also a step towards an intelligent grid (Harris 2008).

Presented at the SCORE! workshop in Paris and at the final conference, the EcoTeams run by Global Action Plan bring representatives from several households together in a group, enabling them to compare and contrast practices among their peers.[10] Global Action Plan's research in influencing consumers also highlights the importance of feeding back information on how changes in behaviour relate to achievement in reducing impacts.

Fischer and Global Action Plan both show that feedback does have an influence; it lowers consumption typically by 5–20%, and that households like it. Feedback is more efficient when users are actively involved in the processes (for example, through meter checking), when there are multiple options for feedback (by time-period, appliance, etc,), when the feedback is frequent (more than monthly) and when it enables comparisons. However, Global Action Plan recognises that participation in the voluntary EcoTeam initiative, for example to facilitate and share experience of saving energy, favours those who are already motivated to modify their behaviour.

In Chapter 7 Kaltenegger and Tisch identified and analysed barriers linked to the implementation of product–service systems (PSSs) in public procurement in Austria and strategies to deal with them in the context of energy-saving performance contracting for federally owned public buildings in Austria. This case shows in concrete ways how large bodies such as the State can have a positive influence at a global level through relatively simple measures if these are implemented in a realistic manner.

Reusswig *et al.* (Chapter 10) analyse the history of wind energy generation in Germany. This case shows how different regulatory frameworks are enabling a gradual structuring of wind energy production and consumption in Germany. At first, consumer cooperative networks, driven by ideology, set up small-scale cooperative structures around local wind turbines. This was enabled by regulation that supports the cost difference from current large-scale (fossil and nuclear) providers. This generates a domestic demand, creates experience and reduced costs. The movement gained enough momentum to influence regulation, a new lifestyle and a socio-technical culture. On this prepared ground, with enough demand for establishing a market, economic regulation can take over. This process culminated in Germany opting out of nuclear energy generation in the 2000s. This raises the obvious point that there can be no economic opt out of the present system if there is no economically sustainable alternative; it also shows that the basis for the infrastructure for an alternative solution can slowly be constructed by local experiments until the alternative system reaches some threshold that convinces market forces at a societal level. The point made by Reusswig *et al.* is that in this process the political, economic, social and lifestyle levels are interconnected in a slow transitional process. However, they note that high transaction costs and problems in motivating people to take collective action means that even stringent disclosure requirements for annual electricity bills will not on its own lead to substantial consumer action. In this context, government action is needed, or a shift in focus, addressing people as citizens and not as consumers. This second approach has led to a doubling of new 'green' contracts since Autumn 2006. This has been helped by the media attention given to climate change.

Still, the market share of 'green energy', although growing, is only reaching 2% in

---

10   For more information on Global Action Plan and its EcoTeams, go to www.globalactionplan.org.uk/ecoteams-0 (accessed September 2010)

Europe. Continuous media pressure and legislative push seem to be necessary to support the recent growth of this market.

At an individual level, the individual economic gains from reducing ongoing fuel bills are a common motivator for consumers to justify the initial cost and to adopt existing low- and zero-carbon (LZC) technologies, such as solar heating, heating controls and low-energy lighting. Government subsidies and fuel prices have therefore helped to tip the balance in favour of adopting LZC technologies for those where disposable income is otherwise a barrier. Taking the UK as an example, reported in Caird *et al.* (2008), adoption of LZC technologies has been slow. Improvements have been mainly incremental, driven by regulation rather than consumer experience. The survey by Caird *et al.* (2008), through the Open University, shows that attitudes and the symbolic value of adopting LZC technologies are also factors in the initial decision to explore LZC options, followed by understanding of how well the technology matches and interconnects with the existing domestic system and whether it operates in a user-friendly manner.

11.2.3.4 Connecting the three observed innovative situations with classic innovation models

Let us try to summarise what we have learned from the cases. These three typical situations are more or less steps to regime change. In Step 1, a few creative agents with a different decision-making system (moral, political, scientific, etc.) generate new alternatives. In Step 2, social participation in local communities increased the individual benefits of motivated agents. In Steps 1 and 2, self selection of agents with atypical motivation or local interests is sufficient to produce niches, but in Step 3, as the scale becomes larger, single agents cannot carry enough influence, and large institutions must be involved, such as states, corporations and professional associations

As we take an evidence-based approach, the situation appears, of course, more complicated than in the models. In practice, it is difficult to bound levels in steps, since local initiatives at the micro-level often build up on meso-level national regulations (e.g. state subsidies, norms) or macro-level trends (media coverage, environmental awareness). For example, the creation of an ESCO by the town of Woking is both an innovation system (an instrument to foster all kinds of new energy services) and system innovation (an attempt to regulate the local energy provision system). Also, even the idea of steps is simplistic, as obviously these 'steps' do not come in clear historical succession; they take place in the same time-frame in different places and at different levels.

All these cases are under the constraint of institutional regulation: corporate, community, government, unions and so on. The role of consumers, business and policy-makers is useful at each of the three, as is the role of the mass media and NGOs.

In a free-market regime, which is generally the case for the consumer economies we consider here, individual behaviour is based on 'free' individual choices. The question is, how can we frame the situation so that individual choices go in 'the right direction'—'nudging' (Thaler and Sunstein, 2008)—so that, when aggregated at a societal level, the result is sustainable? The present situation is precisely an illustration of a system where individual choices (for individual comfort, for local corporate profit, etc.) are problematic at a global level because they are largely unsustainable. If we stay in a free-market regime the transition to sustainability must therefore encourage individual decision-making systems and tools that ensure that individual choices construct sustainable results when aggregated. For example, we need to ensure that consumers will choose 'greener' alternatives for the energy they use at home, that builders will build 'greener' houses, that investors will fund them and so on. Individual agents must believe that, somehow, the choice they make is individually beneficial compared with other alternatives.

The decision system of agents is based on learned habits and enforced regulations and is governed by a social value system. Choice is made from between a limited array of offers. Agents use a series of mental, social and technical tools to evaluate the choice alternatives. All these elements are to be considered when preparing a transition in the regime responding to the potential for change toward SCP.

11.3 Potential for change: the determinants and process

With the practical experience and lessons from the case studies in mind, this section tries to offer a framework to help actors identify their role, opportunities and actions in change towards more sustainable energy use in houses and buildings and the potential obstacles they need to overcome.

First, we describe the general context of stability. In trying to push their agenda, actors confront the mechanisms of social stabilisation, which are designed to maintain existing systems and therefore hinder change. Section 11.3.1 describes what regulatory mechanisms guide and control individual behaviour in stable systems. Not only is it important to understand these mechanisms because they may oppose change but also, as the target of sustainable change is to reach such a stable state, it is worth identifying the main mechanisms regulating the system. From this framework, we shall deduce what constraints and targets should be addressed in attempts to construct a new dynamic equilibrium. This is the triple-determination framework. Section 11.3.2 describes the leopard skin process of change in which these three layers interact, and explains why evolution is a mixture of geographical local change and non-local diffusion. Section 11.3.3 briefly positions in the leopard skin model these three typical situations we have observed in the case studies.

11.3.1 Why do we act as we do? The triple-determination framework of affordances, representations and control institutions

The triple-determination framework of installation theory (Lahlou 2008b) addresses the determinants of individual behaviour. At a given moment, the world can be considered as an 'installation' of distributed features in the physical environment, people and institutions. This installation guides people into their paths of activity (Lahlou 2008a).

The physical level provides affordances (Gibson 1982) for activity: that is, which activities can be supported by physical objects. For example, houses afford shelter, washing machines afford laundering. This is the first level of determination. One can only do what is afforded by the existing environment. In our case, a society can use housing sustainably only if there are sustainable houses available.

At a psychological level, representations provide possible interpretations of the situation and enable subjects to elaborate and plan behaviour. Mere affordances are not sufficient. This is the second level of determination: people can use mental representations to interpret affordances to support their activities. Even if there are sustainable houses, there are many ways of using them in an unsustainable way, such as by over-heating them yet leaving the windows open and so on. An unoccupied room should have the connotation of 'lights must be switched off'.

Representations and physical objects follow a process of co-evolution; representations are constructed by the practices people follow in using objects. Conversely, objects are made after the pattern of their representation; houses are built after the representation of houses. This is why representations match with objects. So if we want new houses to be sustainable, the very idea of a 'house' must become sustainable so the design(er) incorporates sustainability.

This local co-evolution of objects and representations is monitored by local communities of interest in a domain such as housing (users, providers, public authorities, etc.) who set the patterns of objects, the rules of practice and so on, because these stakeholders know the field, objects, representations and rules are adapted to behaviour. These stakeholders create institutions that consist of sets of rules to be applied to keep order, to ensure cooperation and to make communities of interest aware that they are playing the same game.

Indeed, we need 'rules of the game'. Knowing how to use the affordances is not always sufficient to create the desired behaviour. Some people might do wrong and provoke (by ignorance, personal interest, etc.) negative externalities for others. Institutions are the social answer; they create and enforce rules to control these potential misuses or abuses. Many of these rules are already contained in the representations, which are, by their nature, normative. Institutions bring a physical control layer to these norms. They enforce them with special personnel; every member of the community tends to serve as a rule-enforcer and brings 'mavericks' back on track. Often, these rules are made formal and explicit (through regulations, laws, etc.) but they may stay as informal rules of good practice or traditions. As these rules are the result of compromise between local interests, they vary from place to place. One only needs to look at the differences between regional architecture for a tangible illustration.

Institutions provide rules and scripts for behaviour; they enable collaboration. This point is crucial in our complex societies where labour is divided and where interrelations between sectors are complex and ubiquitous. So the institutional level is the third level of determination.

The resources and constraints provided at these three levels guide our social life and make it possible and fluid. Smooth operation of the system is obtained because the design of these three levels of activity support is negotiated between communities of interest on the basis of experience of what is desirable for society. Institutions are the social constructs that emerge in these local ecologies and monitor and rule their evolution. The system is very stable because there is a considerable redundancy between these three types of determinants; therefore even if one fails to be guided by one of these determinants, the others will put back on track the person who 'wanders from the path'.

This triple determination framework explains how we behave at a given moment in time. This works because representations of objects match their actual shape and affordance, and because rules prescribe behaviours that are indeed feasible in the state of the art. Now, as stated earlier, this matching comes from a slow co-evolution between objects and representations. This co-evolution is done under the continuous monitoring and control of stakeholder communities that use institutions as social and economic tools to safeguard their interests. This monitoring explains why the state of things usually reflects some *rapport de force* between communities of interest (for example, between house builders and inhabitants). This is on a par with the fact that institutions are enforcement tools at the service of some community. In trying to push towards a more sustainable society, we try to bias this co-evolution; therefore we have to deal with the existing communities of the domain, take into account existing *rapports de force* and eventually create new ones.

For our problem, changing society into a more sustainable system, and, more modestly, to reduce energy use in housing, we must keep in mind all three levels in order to understand the evolution of the world. Reaching a stable and sustainable state of the system involves making changes at three levels: the physical (to houses and the energy supply within the technical system of consumption); the representational (the ideas people have of what a house is and what living in buildings means); the institutional (the rules of good practice accepted and enforced by stakeholders of the domain). This is especially true for the housing and energy issue, as we have seen that the awareness–action gap is strong (between representations and practice). Therefore, installing affordances that trigger sustainable behaviour is one of the ways of filling this gap.

In historical cultural change, evolution is very slow because it needs distributed changes. Here, in our domain, changes must take place in all the physical systems (millions of individual buildings), in vast energy supply chains, in the attitudes and practice of millions of individual users and in thousands of laws and regulations. This extensive aspect of the real world system is a challenge to general theories because the field displays considerable local variation, thus what may work 'here', may not be easily transferable to 'there'. Also discussed above, there are three intertwined layers. Finally, the field is not static, but rather active, and reactive; many stakeholders will resist change as long as it is not a win situation for them.

As we saw earlier, in the case of housing there is considerable physical inertia because of the slow turnover of the installed buildings. Housing is connected with traditional practice and many cultural constructs. Also, there are many regulations on which changes depend, and these are controlled by very different types of institutions: states, businesses, local social group cultures and so on. So, how can we change energy use in houses and buildings?

11.3.2 The three-layered leopard skin change process

The change process is considerably dependent on local conditions and accidental opportunities. The case studies collected in this book suggest that, at least in the domain of energy use in houses and buildings, changes can follow a 'leopard skin' change process, starting with a series of local 'spots of change'. That is, local experiments create spots of change, which in some cases grow larger at a community level. Possibly, the whole system will finally change when these locally grown experiments merge, or start being common enough to become the new norm. We will show on the basis of these cases how the three-layered framework (physical, representational and institutional) can be applied to this leopard skin.

As a potential actor, or stakeholder in the decision-making process, it may be unclear as to what possible action can be taken. The three-layered leopard skin framework is a way for actors to consider possible strategies for pushing their specific agenda by drawing on the available resources in each of the three layers, depending on their position and local opportunities. What is the current map of the leopard skin in my domain of interest? Where am I, in which kind of spot? How do I have to stabilise the various levels of the skin to make the spot sustainable? How can I spread or replicate the change?

This model differs slightly from simple leopard skin geographic diffusion in that here diffusion does not happen simply by geographical contiguity but also by rebounds through the three layers of the triple-determination model.

Depending on the process with which one is involved, some parts of the global system may be more or less easy to change. At the niche level, one cannot generally change institutions, but one may change the technology and local practice within local experiments. This concerns the activities of a few hundred people at a maximum. This is done mainly on the basis of inter-individual relations and negotiations, hence the importance of individual 'champions' with appropriate communication and negotiation skills.

At the community level one can change some institutions, but not the technology; also, some community-level practice can be institutionalised. Institutionalisation is done through the creation of local rules; so political and administrative proficiency is necessary.

At the market level one can change the practice of millions by bringing in regulation to drive changes in those institutions already leading change in products and behaviour, or by proposing products that afford and demonstrate in their representation widely recognisable benefits within existing institutions. Negotiation is possible with some national institutions, and one can disseminate an existing technology. Economic *rapports de force* are critical. This does not mean that, for example, it is impossible to change technology at the community level, but this may be achieved indirectly and through resourcing local experiments within the community as a niche.

Most complex problems happen at the meso-level because one cannot use the rest of the system as commons (for example, the grid). At the local level, negotiation is at least possible with most local stakeholders, who can be easily identified, and there may be enough independence from the global framework to consider the global framework fixed. At the global level, negotiation is possible with other large stakeholders, but is very difficult with individual end-users.

In fact, especially at the community and market level, successful actors tend to overcome local obstacles by alternating their use of different layers of the leopard skin. For example, when confronted with a dead end in terms of technical development they may draw on institutions to change the framework. For example, where local authorities cannot provide public subsidy to balance the commercial risk of community-level renewable energy, the establishment of an ESCO, as described in Chapter 6 by Thorp for the case of Woking, provides a new institution for energy provision at the community level, compatible with both the new affordance of investment in renewable energy technology and the wider market institutions prescribing commercial return on investment. The German situation described by Reusswig *et al.* in Chapter 10 provides a detailed account of the interplay on a larger scale between the many institutions that populate the leopard skin and how consistent minorities used them to push their case when windows of opportunities opened, therefore enabling the adequate legal and political framework for the local initiative. The reverse may also be true; when institutions block a path, creation of a local physical affordance may solve the issue: this is what Disch did when he created his Solar Funds to bypass the classic banking institutions that did not provide adequate financial support (see the case study presented by Wüstenhagen, in Chapter 5). Representations from the experience of gaining a successful application for the first wind turbine for the Hockerton Housing Project helped the local authority (in consultation with other stakeholders) to consider the benefits and absence of anticipated problems of erecting a wind turbine, enabling them to grant planning permission for a second wind turbine much more quickly.

11.3.3 Typical situations of change, and the leopard skin

As we have suggested in Section 11.2.3 we can distinguish in the leopard skin process three typical situations: local experiments (a champion quest), the innovative community, and the larger, non-local innovation (the transition to a more mainstream market).

In the 'champion's quest', a small spot of innovation is being created in the physical layer of the leopard skin. In this situation, ideologically motivated 'champions', often architects, create a sustainable building with more sustainable energy use built in, at considerable personal effort, by gathering around them a team or network of supporters. At this step, actors concentrate on creating locally new physical affordances (a house, energy provision system and so on). The representations are pre-established by sharing the ideological vision within the project team, and the debate is mostly technical and financial.

In the second step, a local community creates, deliberately and collectively, a local sustainable system, with the active complicity of a few individual activists holding key positions in institutions. A local effort is undertaken to install an innovating system that combines the three layers of the leopard skin in a limited geographical area. At this level again, the role of individual visionaries (aediles) is important, but the community as a group and as a geographical entity is the emerging actor, constrained by the wider institutional inertia and, potentially, by resources to create or adapt physical affordances that can boost the mental representations of what is achievable—towards sustainable energy use in buildings.

In the transition to a wider market situation, professional communities try to create institutions (rules, norms, organisations, large-scale production or regulation systems, etc.) that have global reach. So, this is mostly in the institutional layer. Typically, in this third type, some stakeholders find business models and, based on the market in local communities, manage to lobby at the national level and obtain new regulations that make their 'business' economically sustainable. Business should be taken here more in the sense of general organisational activity, as this business may be non-commercial. There are few examples yet of activity at this level, and it is too early to draw conclusions. The nature of actors here can be companies, institutions, 'lobbies' or sets of actors assembling into communities of interest, such as associations, social movements, professional unions or networks.

There is a final step we have not seen yet, where these businesses or large institutions reverse the whole market and make their own model dominant, or at least prominent, by converting, overcoming or outing the other existing models.

So, in practice, changes do not occur in one 'level' of determination (such as the physical, or representational) described in the triple-determination model, and then in another one. Nor do changes simply disseminate geographically, like a stain of paint would spread out on the skin. Rather, they operate by creating local seeds of change by installing physical innovations (here, sustainable buildings) in the physical layer, then feeding local innovative communities where all three layers (physical, representational and institutional) combine together as a system. Finally, there may be increased 'greening' by dissemination through market mechanisms and non-local change at the institutional level, which gradually makes 'mainstream' transitional change out of local experiments. This institutional layer creates in other places the conditions for possible local spots of change, therefore multiplying the spots of change in a leopard skin pattern. In this process, local actors who are ready for change seize, for the benefit of their own project, the institutional opportunities created by other distant actors. Every little spot of change is slightly different as local stakeholders adapt in an opportunistic and realistic fashion to local opportunities and constraints. As these spots develop and merge, they eventually create a more uniform culture. This process of local adaptation is necessary to produce sustainable systems, where stakeholders negotiate trade-offs in the various dimensions (social, economic, environmental), so governments must be careful to leave some space for variety.

Global change will be the result of a sum of geographically local changes, but the mechanism of dissemination is not geographical, it is social. As we can see, it is at the institutional level that the changes find the most powerful mechanism for replication.

11.4 A short comparison with the mobility and food domains

Alongside this work on energy use in houses and buildings, the SCORE! project provided analyses of two other priority consumption domains: mobility, and agriculture and food. The result of this work has been published in two other books in the System Innovation for Sustainability series (Geerken and Borup 2009; Tischner *et al.* 2010). This section gives a short comparison of the main differences and similarities across the three domains, using some key concepts introduced in the first SCORE! book (Tukker *et al.*, 2008) and elaborated in Chapter 2 and these conclusions.

It concerns the characteristics of the system: is there agreement on the sense of urgency of change and the means to realise change? What are the key players in the system that may drive change? It also concerns the potential role of each actor in the 'triangle of change': what could consumers, producers and governments do? Finally, it asks which factors contribute to windows of opportunity for change and what are the main challenges.

Table 11.1 provides such a comparison, summarising information from Chapter 2 and the concluding chapter of each book in this series. We see here that in the different domains rather different strategies for change have to be chosen, since the three fields have different characteristics.

When it comes down to agreement on the sense of urgency and the means to realise change, the food area appears to be a complicated one. Some would pursue organic and local food production, whereas others point at the fact that precision-farmed food produced at the place where local conditions are optimal can be sustainable given the relatively low impact of transport (for example, by sea). Genetically modified organisms (GMOs) are especially controversial; they are seen as a means to enhance food production sustainably by some but as fully unnatural by others. The sense of urgency in the areas of mobility (energy use, congestion) and housing (energy use) are more commonly shared. Here, the only dispute is to what extent one could pursue an outright reduction of mobility or housing area demands. The housing area is also the domain where the direction toward a solutions seems clear (energy-neutral housing), where the other areas still need more convergence in terms of insight into what approaches might work.

When it comes down to the potential role of actors in the triangle of change (consumers, producers, government), one sees important differences too. Unlike in the case of mobility and housing or energy-using products, in the food area consumers have a relatively high degree of freedom in terms of consumer choice. (In the area of housing and energy use consumers in most EU countries now also have a free choice of energy suppliers, including those offering a proportion of renewable energy up to 100%, but the practical hurdles in this field seem still significant.) Although they are obviously constrained by habits, routines and so on consumers can easily switch food choices and even diets in rather radical ways. Such choices can be made on an almost daily basis, whereas when it comes to mobility and energy use in housing, the physical lock-ins significantly reduce consumer choice.

Compared with the building sector, it seems that the power on the production side in mobility (held mainly by car and infrastructure providers) and agriculture and food (held mainly by retailers, major food producers) is probably much more concentrated. Interestingly, in the food area one sees that such mainstream players seem more active in pursuing sustainability goals and related management systems than in the car industry, most probably because of the greater flexibility and implicit consumer expectations with regard to environmental and social responsibility. The driving forces, main challenges and the role of government also seem rather different in each domain.

- In the building area the main challenge seems to be to mainstream solutions, focusing on realising zero-energy houses. For new houses, this implies mainstreaming proven solutions, whereas for existing housing stock the question on how best to tackle the problem is still open. A smart innovation and diffusion policy sufficiently flexible to be adopted and implemented locally seems the main challenge here

- In the food area, one sees an implicit societal negotiation going on about what implies 'sustainable food'. There seems, however, to be a relatively high degree of flexibility in the food system, and a supportive government role in the background may be sufficient

- In the mobility area, the problems and lock-ins seem most problematic. Given the important role of infrastructure, spatial planning, the need to provide alternative transport modes and the fact that the few success stories rely heavily on government incentives,[11] the need for a strong integral government role seems inevitable

TABLE 11.1 Comparison of food, mobility and housing domains; system, triangle of change, factors providing windows of opportunity and main challenges

Food		Mobility		Housing	
Score	Comment	Score	Comment	Score	Comment
(a) The system					
Agreement on goals and sense of urgency:					
Low	There is no full consensus on what a sustainable food system will look like. There is dispute about the sustainability benefit of GMOs versus regular food, organic versus precision-farmed food, and locally grown versus imported food	Medium	Reducing congestion, CO_2 emissions and emissions of other pollutants are generally accepted goals. The need for an absolute reduction of mobility is much less so	High	Zero-energy houses and low-energy products are seen as goals. Limiting the housing area or energy consumed per person is not, however, feasible

---

11 See, for instance, the case of the London congestion tax, discussed in the book on mobility (Geerken and Borup 2009: Chapter 8).

Food		Mobility		Housing	
Score	Comment	Score	Comment	Score	Comment
Certainty or agreement about means:					
Low	As above	Low	There are few success stories of change; the growth of mobility demand seems unstoppable	Medium	There are many successful illustrative projects on zero-energy houses available, but mainstreaming seems difficult
Geographical characteristics:					
Global	Most modern food chains span the globe	Global and local	There is just a handful car manufacturers left. However, the local situation highly determines mobility patterns and the availability of alternatives	Local	The building industry is rather locally organised and has local characteristics
Power nodes in the system:					
	Retailers and major food companies		Car producers and infrastructure providers		Varies by country; social housing agencies, landowners, developers, builders and refurbishers, and customers

(b) The triangle of change

Food		Mobility		Housing	
Potential for consumer-driven change:					
Medium	Consumers make daily purchase choices and can change diets, though in practice they are locked in by routines, habits and availability of food alternatives	Low	Consumers have often limited choice with regard to their pattern and means of mobility, except for during holidays	Low	In most countries consumers have little power over how houses are built

Food		Mobility		Housing	
Score	Comment	Score	Comment	Score	Comment
Potential for producer-driven change:					
Medium to high	Retailers (e.g. choice editing), major food companies (e.g. via certification schemes such as MSC) and smaller companies (e.g. Bionade) all have proven to be able to drive change	Low to medium	Automotive companies at best will realise incremental change; alternative mobility providers tend to stay in niches	Medium	Many illustration projects are available, but mainstreaming seems difficult
Potential for government-driven change:					
Medium	Government can support change by public procurement and regulation, but regulation will work mainly to set minimum standards rather than to drive radical change	Medium to high	Government has a major role in spatial planning, the availability of public transport and encouraging a more sustainable modal split	Medium	Via energy labelling and minimum standards governments can ensure a basic energy performance, but radical change is possible only if zero-energy housing is mainstreamed
(c) Factors providing windows of opportunity					
	Consumer expectations about health, safety and social issue; reformulated CAP		High energy prices; congestion		High energy prices
(d) Main challenges					
	Negotiating a view on sustainable food and implementing this. Various views and related production structures may co-exist		Overcoming various lock-ins (regional planning, physical infrastructure, the importance of the car industry to many industrialised countries)		Mainstreaming of proven practices with regard to low-energy houses Dealing with the existing housing stock

GMO = genetically modified organism   MSC = Marine Stewardship Council   CAP = Common Agricultural Policy
CO_2 = carbon dioxide

11.5 Issues and recommendations for action

The actors involved in the change processes are many. Chapter 2 described some of them: builders, architects, ESCOs, inhabitants, local authorities, state agencies and regulators, energy technology suppliers, building technology and service suppliers, developers, housing associations and landlords, financiers and so on. Rather than proposing a systematic grid that includes all potential actors within a time-frame this section focuses on some critical issues for success in typical situations representing the three steps to transition highlighted in the previous sections. We also profile the types of change commonly considered to improve energy use in housing as well as barriers to change, suggesting some policy and actor roles.

These typical situations may not cover all cases; furthermore they are dynamic, and every local case is different. That said, it is always interesting to see what problems others have encountered in similar situations and how they overcame those problems. We hope this list will help actors through the difficult, and multi-layered, process of change according to their situation, local, community or non-local.

11.5.1 Supporting progress through three steps of transition

11.5.1.1 Issues and recommendations in a local experiment situation

This situation should be considered not only from the champion's perspective, but also from the perspective of the other stakeholders who are interested in the success of the local experiment.

Champion-led innovation needs a good champion, driven by a vision, deeply invested in the project, charismatic, creative and with hands-on technical expertise. She or he is often a technology fan. Success factors are that the champion:

- Heads the project, manages to raise and secure funds along project time-lines, including for unexpected demands (for example, she or he may have powerful sponsors in funding bodies)

- Adapts in a pragmatic ways to circumstances and creates alliances with local stakeholders

- Has good local relationships with the 'bounded socio-technical experiment' (BSTE) site

- Stimulates enough internal support in the project team, based on shared values

Strong motivation of the other participants is also a success factor.

There are, of course, technical issues connected with innovation. Experience shows that, given sufficient finance, these issues are usually solved, at the cost of delaying the general project.

Financial issues are critical. Such projects are investment projects, with rather long-term return on investment. Funding that does not require a high rate of return needs to be secured. Financial capacity from mixing traditional funding with specific resources (such as grants, public subsidies, corporate social responsibility [CSR] funding in companies, etc.) should be considered in the design of a specific project investment plan, and especially in the timing of the delivery of those funds. The capacity to generate stable return on investment in the long term should be a priority. The (improved) asset value of a building with an upgraded specification may help balance the investment required.

As these innovative projects are usually in competition with classic projects, regional stakeholders need to take care that the terms of competition do not add excessive overheads to innovative projects; for example, the terms used in calls for tender specifications requiring only an appropriate burden of proof of compatibility with existing norms or procedures. Otherwise, the existing terms of competition could perhaps prescribe unsustainable solutions.

Low reliability of suppliers of new materials and equipment is a risk that must be covered in the contracts or resources for contingencies.

Traditionalism and risk aversion of public authorities and financial institutions is an obstacle. When these institutions bring support, this is often at the initiative of some visionary sponsor inside the institution. Therefore a turnover of people sponsoring the experiment may be fatal. This may be especially true if funding relies on an individual's personal valuation not reflected in the wider priorities of the sponsoring organisation. Innovation sponsors should ensure the stability of allocated funds in the long run.

Media and political support, or at least 'drive' from some local influential stakeholder, may be based on the innovative and ideological aspects.

The possible value from creating a 'free' inherent resource, for example for demonstration or communication purposes, can help to reach an economic balance.

11.5.1.2 Issues and recommendations in a community innovation situation

As Ürge-Vorsatz and Miladinova (2005: 261) note for the new Eastern European EU member states, 'since many efforts contributing to the sustainability of the energy sector are rooted at the community level, regional-, municipal- and community-based initiatives should be encouraged more'. At this level, local authorities have a crucial role to play, but of course they cannot operate alone, and the main challenge is to organise a local network of operators, and their relations with their local clients.

Low-rate, long-term return on investment can be obtained on such projects, but they often do not appeal to classic commercial operators—it is crucial to find the appropriate funding structure. It is good practice for funding of these projects be done on a separate account, and preferably through an institution separate from local authorities. Buildings and their associated organisations should be economically sustainable in the long run. Such structures can be created if needed (cf. the ESCO in the Woking case presented by Thorp in Chapter 6).

Decoupling the political agenda from the functional agenda of the project is also a healthy move. The Woking case is a remarkable example of a solution for this. Turnover in personnel (through political cycles, or normal churn) in the local authorities can be a major source of failure if the delimiting and autonomy of the new system has not been clearly organised from the start.

In practice, innovative communities often combine several 'local experiments' (buildings, generation systems, funding schemes, etc.). For each sub-project, the success conditions described in Section 11.5.1.1 apply. Making the whole network of actors work in a local subsystem and within a wider system introduces systemic issues with specific technical problems, struggles for shared representations and transaction costs. For example, technical issues related to grid connection may jeopardise the whole system. These emerging sub-projects should be identified and dealt with by specific resources and personnel and not simply considered as new issues between existing, inexperienced stakeholders, in order to avoid local tragedies of commons.

Enrolling the stakeholders in the new local regime is easier if they are represented through institutions and if the relations between stakeholders can take place through direct economic transactions. This both empowers the actors and helps distinguish the sociotechnical issues from the political and governance issues.

Competition with classic projects is a continuous and implicit challenge. Traditionalism and risk aversion in public authorities and financial institutions are common. Any crisis, conflict of interest or dispute may raise the question of whether to continue with the innovation process. Therefore monitoring and evaluation systems, including financial evaluation, should be set up from the start in order to enable fair and informed discussion in these moments of tension.

11.5.1.3 Issues and recommendations in a non-local situation

It is more difficult to provide critical success issues in this very large and diverse situation. In contrast to the other, local, situations, the arena is in these cases very open, and one of the issues is precisely to get in touch with the right stakeholders. In this section we provide some examples of strategies.

11.5.1.3.1 Pursuing change through government regulation

Government regulation might be measures providing incentives or restrictions. In general, a mild approach is adopted, for example acting through market mechanisms such as labelling, instead of forcing stakeholders to change practices or technology, as was done to ban the use of chlorofluorocarbon (CFC) gases through the Montreal Protocol.

Regulation by market forces raises two issues in response to demands to consume and the pressure for growth:

● The presence of less government control over the outcomes is a limitation

● It is easier to develop local initiatives to match a variety of (external) funding sources and administrative paths to help installation

Some institutional strategies set up in Europe have been based on direct financial help such as through subsidised solar generation or building insulation, with the provision of tax credits or other financial instruments. These are useful to start a market hampered by lack of funding for technological breakthroughs but may be counter-effective if subsidies distort the market in the longer term.

Another approach has been to make the cost of unsustainable energy use more visible, for example through the Emissions Trading Scheme. In the same vein, the EPBD and Directive on the Indication by Labelling and Standard Product Information of the Consumption of Energy and Other Resources by Household Appliances (European Council 1992, 2003) make compulsory the standardised measurement and display of energy efficiency of homes and appliances. This offers incentives for stakeholders to modifying their products to compete with those displaying better energy performance.

Green procurement policy is also a means of securing a market of some scale and continuity for greener technology and service suppliers.

General recommendations include the need to signal regulations accurately and sufficiently in advance of implementation to allow product development and supply adjustments to be made. A further principle, seen in the implementation of the UK Code for Sustainable Homes (UK DCLG 2009), is the specification of levels of performance (against standardised metrics) that can then be used to progressively offer a recognisable scale for high advisory levels and low mandatory levels, facilitating the introduction of subsequent increases on the existing, recognised scale.

11.5.1.3.2 Pursuing change through large corporate initiatives

Large corporations do have a major impact on energy use in housing regime. Demand-side control energy use is interesting for electricity companies because it helps to lower peak load and therefore minimises generation installation and network capacity. Some operators (such as EDF in France) started it as early as 1965.

Green contracts, where the renewable origin of energy is guaranteed, can be proposed by energy companies to fund their green investment and can be favoured through green provisioning.

Professionals can make and enforce codes of good practice, in order to green the rules of the game; presently, those who are trying to foster sustainable products at the expense of lower or delayed profit are penalised in competition.

Consumer feedback experiments (such as those described by Fischer in Chapter 9) should be pursued to inform consumers, who will otherwise remain almost unaware of the relation between their electricity-consuming behaviour and their utility bill (Yamamoto et al. 2008).

11.5.1.3.3 Pursuing change through professional initiatives and tools

There exist widely applicable and acceptable tools presenting results responsive to conventional as well as sustainable priorities for decision-making, such as BIDS™ (Loftness et al., Chapter 8). The US Green Building Council's Leadership in Energy and Environmental Design (LEED) Green Building Rating System™[12] and US International Energy

---

12 LEED is a third-party certification programme and the US-wide accepted benchmark for the design, construction and operation of high-performance green buildings, recognising performance in five areas: sustainable site development, water savings, energy efficiency, materials selection and indoor environmental quality (www.usgbc.org/DisplayPage.aspx?CategoryID=19 [accessed September 2010]).

Conservation Code (ICC 2004) Compliance Tools[13] are examples of what professional institutions can do.

Feedback systems influencing consumer choice such as described by Fischer (Chapter 9) and the Global Action Plan,[14] or decision systems for professionals such as described by Loftness *et al.* (Chapter 8) are typically the kind of decision tools that can encourage the decisions of individual agents towards more sustainable choices by providing them with evidence showing that these choices are individually beneficial over a longer time-frame.

11.5.1.3.4 Pursuing change through consumer associations and manufacturers' alliances

Consumers associations and manufacturers' alliances and associations can act as intermediaries to analyse, validate and support change. Some are experienced in proposing options liable to be practical from the perspective of their members' interests as well as in lobbying regulators. In addition, consumer associations and manufacturers' alliances and associations could potentially collaborate in areas of common interest. Each provides a forum where individual consumer or manufacturer views can be supplemented by wider experience or consultation.

We suggest the triple determination approach may be of use not only in the typical three situations described here but also more generally (as every situation is a specific, complex and dynamic mix of these prototypes); each actor should consider local conditions and see, in an opportunistic and realistic manner, which of the layers she or he can possibly influence. The key to creating good systems cannot be merely a mechanical application of a set of guidelines; rather, it should consist of the creation of a negotiation arena where stakeholders gradually construct sustainable compromises that can then be 'installed' in the real world through institutions, physical artefacts, representations and practice.

11.6 Conclusions

Given most modern human activity takes place indoors, the societal representations as well as the physical affordances of buildings as they are currently constructed also accounts for many difficulties, both in the present and in the future. Energy efficiency is still a low priority in building design; efficient retrofitting is costly and difficult—representing challenges alongside the slow evolution of more sustainable building practice and parts, especially for HVAC and other infrastructure.

A vast and complex set of rules, institutions, vested interests (including capital) and habits that are the result of historical evolutions and compromises between stakeholders make every change a complex negotiation process, where each player has limited power to drive radical change.

---

13 For more information, see www.energycodes.gov/compliance_tools.stm (accessed October 2009).
14 See footnote 10.

As new construction is marginal in our economies (unlike in China[15]) there is no rapid growth of infrastructure offing change in the regime by outnumbering the present stock of real estate with new energy-efficient buildings. Therefore, changes take place mostly through modification of existing infrastructure (by building retrofitting, adapting the structure of energy provision, changing stable habits, etc.) by using market mechanisms and trying to orient them with state regulation.

Innovation in market economies relies on individual initiative, but it is unclear here what benefit most stakeholders in the domain would value from more sustainable energy use in buildings. Clearly, end-users would get a lower energy bill, but most actors and decision-makers in the domain have little immediate incentive to engage in an uphill green path towards sustainability against the domain's rigidity and inertia.

There is a vast action–awareness gap between what consumers would be like to do for sustainability and what they actually do on an everyday basis. This gap is well documented and the obstacles are known. Some paths of improvement are to provide feedback on actual detailed consumption, as shown in the cases presented here, and making available a wider set of offers through a network of skilled salespeople and installers. The growing awareness of climate change is pushing things in the right direction.

To compensate for the potential loss of comfort and the stronger constraints coming from consuming less energy, we suggest an insistence on the social benefits of participating in local communities sharing resources or services in a sustainable way (humans are social animals by nature): this can be organised at a town level.

We have mentioned some of the major trends currently influencing change towards SCP in energy use in houses and buildings: zero-energy buildings, better energy efficiency in products and practices, product–service systems, demand-side management, intelligent grids and green provisioning. These can take place only if there is substantial space for discussion between stakeholders. Large players (political, economic) still have an important role to play because decisions need to anticipate benefits, some apparent only in the medium and longer term.

Many situations are blocked because stakeholders feel they are locked-in, especially in the step from small- scale innovation to larger, more mainstream implementation. We have proposed a three-layered framework that can be a strategic resource for actors, since they may, when they are blocked by one layer, try to use alternate ways by drawing on another layer.

We have shown how, in sustainable systems, individual and collective behaviour follows a triple determination system: by the affordances of physical objects and installations, by the representations and practices incorporated in individuals' interaction with them and by rules negotiated between stakeholders, then guaranteed and controlled by institutions. Changing the world is a matter of making a new installation that will guide behaviour, and such installation must be distributed in the real world at these three levels: physical installation, representations and practices, and institutions.

We have shown that in the domain of housing there are typically three situations

---

15   The Chinese Ministry of Construction estimates that, between 2005 and 2020, 30 billion m²—half of the world's new construction—will be constructed in China (Ma and Bao 2006, as cited by Yang and Kohler 2008).

where change takes place: local experiments driven by motivated champions, community projects and larger, non-local, change. This is slightly different from a classic 'small–medium–large' change, because change is not a matter of expanding the size of innovation but rather of installing the three layers of behavioural control (physical affordances, social representations, institutions), each layer having its own diffusion patterns and mechanisms.

We have explained how successful change in the domain can follow a leopard skin pattern. Small spots of change in the physical environment (buildings, energy systems, etc.) emerge with the help of motivated individual 'champions' when local conditions are favourable. Some communities manage to create local systems where the three layers (physical, representations, institutions) are coherent with a sustainable lifestyle in a limited geographical area. Large bodies (political, professional unions, etc.), inspired by local initiatives, create larger, non-local institutions (for example, through regulation) that create conditions of possibility for more local change. Then with these better conditions, more local actors seize the opportunity to realise their own projects: new spots of change emerge, and grow, greening the three-layered leopard skin.

Finally, the case studies presented in this book show that, even though the regime is complex and rigid, positive change can and does take place. Another world is possible where steps to SCP are available everywhere for those who see the potential.

References

ASHRAE/IESNA (American Society of Heating, Refrigerating and Air-conditioning Engineers/Illuminating Engineering Society of North America) (2004) 'Standards 90.1-2004: Energy Standards for Buildings except Low-Rise Residential Buildings', ASHRAE; www.ashrae.org/education/page/1834 (accessed September 2010).

Caird, S., R. Roy and H. Herring, H. (2008) 'Improving the Energy Performance of UK Households: Results from Surveys of Consumer Adoption and Use of Low- and Zero-carbon Technologies', *Energy Efficiency* 1.2 (May 2008): 149-66.

ETP (European Technology Platform) (2009) 'SmartGrids: Strategic Deployment Document for Europe's Electricity Networks of the Future'; www.smartgrids.eu (accessed October 2009).

European Council (1992) 'Council Directive 92/75/EEC of 22 September 1992 on the Indication by Labelling and Standard Product Information of the Consumption of Energy and Other Resources by Household Appliances', European Council, European Parliament, Brussels; eur-lex.europa.eu/LexUriServ/LexUriServ.do?uri=CELEX:31992L0075:EN:NOT (accessed October 2009).

—— (2003) 'Energy Performance in Buildings Directive (2002/91/EC) of 4 January 2003', European Council, European Parliament, Brussels; www.euroace.org/reports/CIBSE_EUBD.pdf (accessed October 2009).

—— (2006) 'Directive 2006/32/EC of the European Parliament and of the Council of 5 April 2006 on Energy End-use Efficiency and Energy Services', European Council, European Parliament, Brussels; eur-lex.europa.eu/LexUriServ/LexUriServ.do?uri=OJ:L:2006:114:0064:0064:en:pdf (accessed September 2010).

—— (2009) 'Directive 2009/125/EC of the European Parliament and of the Council of 21 October 2009 establishing a framework for the setting of ecodesign requirements for energy-related products' (Brussels: European Council, European Parliament; eur-lex.europa.eu/LexUriServ/LexUriServ. do?uri=CELEX:32009L0125:EN:NOT [accessed September 2010]).

European Parliament and European Council (2002) 'Directive 2002/91/EC of the European Parliament and of the Council of 16 December 2002 on the Energy Performance of Buildings'; eur-lex.europa. eu/LexUriServ/LexUriServ.do?uri=CELEX:32002L0091:EN:NOT (accessed October 2009).

Eymard-Duvernay, F., and L. Thevenot (1983) *Les investissements de forme: leurs usages pour la main-d'oeuvre* (Paris: National Institute for Statistics and Economic Studies [INSEE]).

Geerken, T., and M. Borup (eds.) (2009) *System Innovation for Sustainability 2: Case Studies in Sustainable Consumption and Production—Mobility* (Sheffield, UK: Greenleaf Publishing).

Gibson, J.J. (1982) 'Notes on Affordances', in E . Reed and R. Jones (eds.), *Reasons for Realism: Selected Essays of James J. Gibson* (London: Lawrence Erlbaum Associates, first published 1967): 401-18.

Harris, E. (2008) 'Upgrading the Grid', *Nature* 454 (31 July 2008): 570-73.

ICC (International Code Council) (2004) International Energy Conservation Code 2004: Energy Standard for Low-rise Residential Buildings set by the International Code Council (Washington, DC: ICC).

IEA/OECD (International Energy Agency/Organisation for Economic Cooperation and Development) (2003) *Energy Efficiency in Economies in Transition: A Policy Priority* (Paris: OECD).

Lahlou, S. (2008a) 'L'Installation du Monde. De la représentation à l'activité en situation', HDR Psychologie, Université de Provence, October 2008; tel.archives-ouvertes.fr/tel-00515114/en (accessed September 2010).

—— (2008b) 'Cognitive Technologies, Social Science and the Three-layered Leopard Skin of Change', *Social Science Information* 47.3: 299-332.

—— and S. Emmert (eds.) (2007) 'Sustainable Consumption and Production Cases in the Domain of Food, Mobility and Housing', *Proceedings of the Sustainable Consumption Research Exchange (SCORE) Network*, Paris, 4–5 June 2007; www.score-network.org/files/9594_Proceedings_worshop.07.pdf (accessed September 2010).

——, J. Maffre and P. Moati (1991*) Régulation des marchés culturels: le rôle de la passion* (Rapport au Ministère de la Culture; August 1991; Paris: Centre de Recherche pour l'Étude et l'Observation des Conditions de Vie [Crédoc]).

Laustsen, J. (2008) 'Energy Efficiency Requirements in Building Codes: Energy Efficiency Policies for New Buildings' (IEA Information Paper, IEA/OECD; International Energy Agency/Organisation for Economic Cooperation and Development, March 2008).

Ma, Y., and S.M. Bao (2006) 'Status Quo of Building Energy Conservation and Green Building in China', *China Construction Newspaper*, 29 March 2006: P1.

Maldonado, E., and E. Oliveira Fernandes (1993) 'Building Thermal Regulations: Why has Summer been Forgotten?', in T. Herzog, N. Kaiser and M. Volz (eds.), *Solar Energy in Architecture and Urban Planning* (Bedford, UK: H.S. Stephens): 626- 30.

Thaler, R.H., and C.R. Sunstein (2008) *Nudge: Improving Decisions about Health, Wealth, and Happiness* (New Haven, CT: Yale University Press).

Tischner, U., E. Stø, U. Kjærnes and A. Tukker (eds.) (2010) *System Innovation for Sustainability 3: Case Studies in Sustainable Consumption and Production—Food and Agriculture* (Sheffield, UK: Greenleaf Publishing).

Tukker, A., M. Charter, C. Vezzoli, E. Stø and M. Munch Andersen (eds.) (2008) *System Innovation for Sustainability 1: Perspectives on Radical Change to Sustainable Consumption and Production* (Sheffield, UK: Greenleaf Publishing).

UK DCLG (Department for Communities and Local Government) (2009) 'Code for Sustainable Homes Technical Guide' (London: Communities and Local Government Publications; www.planningportal. gov.uk/uploads/code_for_sustainable_homes_techguide.pdf [accessed September 2010]).

Ürge-Vorsatz, D., and G. Miladinova (2005) 'Energy Efficiency Policy in an Enlarged European Union: The Eastern Perspective', in *ACEEE 2005 Summer Study: What Works and Who Delivers?* (Washington, DC: American Council for an Energy Efficient Economy [ACEEE]): 253-65.

Von Uexküll, J. (1956) *Streifzüge durch die Umwelten von Tieren und Menschen: Bedeutungslehre* (Hamburg, Germany: Rowohlt Verlag).

—— (1965) *Mondes animaux et monde humain* (translated from the German; Paris: Denoël [1st published 1956]).

WBCSD (World Business Council for Sustainable Development) (2006) Dongtan: The World's First Eco-city (Geneva: WBCSD; www.wbcsd.org/Plugins/DocSearch/details.asp?DocTypeId=251&ObjectId=MTk4MTk [accessed September 2010]).

Yamamoto, Y., A. Suzuki, Y. Fuwa and T. Sato (2008) 'Decision-making in Electrical Appliance Use in the Home', *Energy Policy* 36: 1,679-86.

Yang, W., and N. Kohler (2008) 'Simulation of the Evolution of the Chinese Building and Infrastructure Stock', *Building Research and Information* 36.1: 1-19.

About the contributors

Azizan Aziz is a Senior Research Architect at the Center for Building Performance and Diagnostics, Carnegie Mellon University. His research focuses on the design of integrated systems for commercial buildings, workplace productivity and sustainable zero-energy buildings. Currently Azizan is working on the Building as PowerPlant/Invention Works (BAPP) initiative and the Workplace 20|20/National Environmental Assessment Toolkit (NEAT) project. Azizan teaches Integrated Product Design and LEED and Sustainable Buildings.

Halina Szejnwald Brown is Professor of Environmental Science and Policy at Clark University, Worcester, Massachusetts. She currently works in the areas of environmental policy, sustainable production and consumption, socio-technical transitions, and the role of the corporate sector in sustainability transition. She is the author of two books and over 50 scholarly articles ranging in topics from risk assessment, US and comparative environmental policy, ethics and value issues in international transfer of technology, institutional theory, small-scale socio-technical experiments, socio-technical transitions, and others. Brown received a PhD degree in chemistry from New York University and prior to joining Clark was a chief scientist for the Massachusetts Department of Environmental Protection. Her research included the US, Eastern Europe and Far East Asia. Brown is a Fellow of the International Society for Risk Analysis and Fellow of the American Association for the Advancement of Science. Currently she co-chairs the Citizens' Energy Commission, which advices the mayor of her home city of Newton, Massachusetts.

Martin Charter is the Director of The Centre for Sustainable Design at the University for the Creative Arts (UCA) and a former Visiting Professor of Sustainable Product Design at UCA prior to joining full-time. Since 1988, he has worked at director level in 'business and environment' issues in consultancy, leisure, publishing, training, events and research. He is the author and editor of various books and publications including *Greener Marketing* (Greenleaf Publishing,1992 and 1999), *The Green Management Gurus* (ebook, 1996), *Managing Eco-design* (The Centre for Sustainable Design, 1997) and *Sustainable Solutions* (Greenleaf Publishing, 2001). Martin has an MBA from Aston Business School in the UK, and has interests in sustainable product design, green(er) marketing, and creativity and innovation.

Joonho Choi is a doctoral student in architecture and a researcher in the Robert L. Preger Intelligent Workplace (IW), Carnegie Mellon University, USA.

Dr **Corinna Fischer** is a senior researcher at the Institute for Applied Ecology, Germany, in the field of sustainable consumption. After her training in political science and psychology, she has been working at Deutsches Hygiene-Museum Dresden, at the Free University of Berlin, at the German National Federation of Consumer Organisations, and as a freelance researcher and consultant on sustainable consumption issues with a special focus on energy and climate.

Doris Fuchs is Professor of International Relations and Development at the Westphalian Wilhelms University, Münster, Germany. Her primary areas of research are private governance, sustainable development, food politics and policy, and corporate structural and discursive power. Among her publications are *Business Power in Global Governance* and *An Institutional Basis for Environmental Stewardship*, as well as articles in peer-reviewed journals such as *Millennium*, *Global Environmental Politics*, *Business and Politics*, *International Interactions*, the *Journal of Consumer Policy* and *Energy Policy*.

Since 1972, Professor **Volker Hartkopf** has been teaching and conducting research at Carnegie Mellon University, USA. His work covers a broad range of activities: international initiatives, funded research and professional consulting on building systems integration, advanced technology, building performance, energy conservation, urban revitalisation, third-world housing and disaster prevention. He is recognised as architect on building projects in Germany, Bangladesh, Peru and the United States. An award-winning professor and a frequent keynote speaker in Australia, Europe, Asia, Middle East and the Americas, he has authored over 100 technical publications. He continues his consulting with such organisations as DaimlerChrysler, Volkswagen, Thyssen Krupp, Electricité de France, US Department of State, US Department of Energy, Siemens, among others, and serves on many boards, juries and panels in the United States and abroad. Currently, Dr Hartkopf is serving as Chair of the United Nations Environmental Programme's Sustainable Buildings and Climate Initiative (UNEP SBCI) Think Tank and is the Bayer Material Science EcoCommercial Building Program's Academic Advisor.

Ingrid Kaltenegger studied Chemistry and Environmental Sciences at the Karl Franzens University in Graz. After her studies, Ms Kaltenegger worked as a project manager in the field of cleaner production for three years before she began work at JOANNEUM RESEARCH in 2001. Since then she has been working as a project manager in many projects dealing with product–service systems, corporate networks and corporate social responsibility. She has also been a lecturer on environmental management systems at the ECOPROFIT Academy Graz and the University of Applied Sciences for Management and Business in Vienna and is a certified trainer in Adult Education.

Myung-Joo Kang has worked at GrAT (Center for appropriate Technology) as a Research Fellow since 2005. She obtained her doctoral degree in the field of product–service systems (PSS) with focus on methodological development and applications in industries. Along her studies and professional experiences in Industrial Design (BSc, Korea), Sustainable Product Design (MA, UK), and Human Interface Design (LG Electronics, Korea), she has gathered knowledge and insights on system design for sustainability. Currently she is conducting a number of international projects for Design for Green Growth and Appropriate Technology.

Professor **Saadi Lahlou** (ENSAE, PhD, HDR) is currently director of the Institute of Social Psychology at the London School of Economics and Political Science (UK). He spent many years in industry, recently at Electricité de France as strategic adviser to the Director of R&D; where he founded in 2000 the Laboratory of Design for Cognition (a large living laboratory to develop office work environments), previously as the Director of the Consumer Research Department of the Research Centre for Lifestyles and Social Policies (Crédoc, Paris). Lahlou as a specialist of behaviour and consumer science developed several innovative techniques for activity observation, including the SubCam technique—a head-worn miniature videocamera providing a first-person-perspective detailed recording. His Installation Theory provides a strategic framework for societal change management to innovators, decision-makers and governments.

Vivian Loftness, FAIA, LEEDAP, is a university professor at Carnegie Mellon University, USA, and served for ten years as Head of the School of Architecture. She is an internationally renowned researcher, author and educator in environmental design and sustainability, the integration of advanced building systems, climate and regionalism in architecture, as well as design for health and productivity. Professor Loftness is a key contributor to the development of the Intelligent Workplace—a living laboratory of commercial building innovations for performance, along with authoring a range of publications on international advances in the workplace. She has served on eight National Academy of Science panels as well as being a member of the Academy's Board on Infrastructure and the Constructed Environment, and given three Congressional testimonies on sustainable design. Her work has influenced national policy and building projects, including the Adaptable Workplace Lab at the U.S. General Services Administration and the Laboratory for Cognition at Electricity de France.

Sylvia Lorek has been working as a researcher and policy consultant for sustainable consumption since 1993. Based on her studies in household economics and nutrition (oeco-trophologie) and economics, she combines the individual micro-economic and the societal macroeconomic perspective in her analysis of (un-)sustainable consumption patterns. She heads the Sustainable Europe Research Institute in Germany and works as a lecturer at the University of Applied Sciences in Münster.

Dr **Fritz Reusswig** is a sociologist in Research Domain IV, 'Transdisciplinary Concepts and Methods', at the Potsdam Institute for Climate Impact Research, Germany. He works chiefly on lifestyle and consumption issues as drivers for global environmental change, especially climate change. In addition, he is interested in the role of lifestyle and consumption changes for a system-wide sustainability transition. Fritz joined PIK in 1995, coming from Frankfurt University. In 2008, Fritz completed his habilitation at the Wiso Faculty at Potsdam University; his habilitation thesis 'Consuming Nature: Modern Lifestyles and their Environment' is prepared for publication. Fritz currently teaches sociology at Potsdam University, at the Brandenburgische Technische University (BTU) in Cottbus and at the Hochschule für Gestaltung in Offenbach.

Megan Snyder is a PhD Candidate at the Centre for Building Performance and Diagnostics, Carnegie Mellon University, USA.

John P. Thorp is Group Managing Director of the Thameswey Group of Companies in Woking, UK. Thameswey Ltd (TW) is an energy and environmental services company (EESCO) which enters into public–private joint ventures throughout the UK to deliver energy and environmental strategies and targets. He gained a BSc(Hons) in 1974, followed by several years in Papua New Guinea in research and marine policy development. With an MBA from Cranfield University, he has held senior posts with many organisations including the International Maritime Organisation. John is a past Chair of the International Solar Energy Society in the United Kingdom (UK-ISES). He is a Chartered biologist, a Member of the Society of Biologists, a Fellow of the Energy Institute, a Fellow of the Royal Society of Arts and a Freeman of the City of London.

Angelika Tisch studied technical environmental protection at the Technical University in Berlin (TUB). She completed her thesis in 2002 at the Institute of Process Engineering at TUB, then worked there as a research scientist in the field of 'Social-Ecological Research/Gender and Environment'. Since April 2006, Angelika Tisch has been a scientist in the department of Ecological Product Policy at the Interuniversity Research Center for Technology, Work and Culture. Her main fields of activity are green procurement, product–service systems, and methods for the ecological assessment of products.

Dr **Arnold Tukker** joined TNO in 1990 after some time working for the Dutch Environment Ministry. Over time, his focus shifted from life-cycle assessment, material flow analysis and risk assessment to interactive policy-making and sustainable system innovation and transition management. In 1998 he published a book on societal disputes on toxic substances, for which he was awarded a PhD from Tilburg University. He has published about 40 peer-reviewed papers, 5 books, 10 book chapters and 150 other publications, and is frequently asked as invited speaker worldwide. In his career, he has been awarded over €15 million in mainly international research grants. He currently manages the research programme on Transitions and System Innovation within TNO Built Environment and Geosciences, Business Unit Innovation and Environment. This programme was evaluated as one of TNO's top-ranking programmes during the 2006 scientific assessment exercise. Arnold is the initiator and manager of the SCORE! network. From 1 April 2010, Arnold is also part-time professor of sustainable innovation at the Industrial Ecology Program/Department of Product Design, NTNU, Trondheim, Norway.

Philip J. Vergragt is a Professor Emeritus of Technology Assessment at Delft University of Technology in the Netherlands. He is currently a Senior Associate at Tellus Institute in Boston, USA, and a Research Professor at Clark University, Worcester, MA, USA. Before moving to the USA in 2003, he also was a Deputy Director of the Dutch government's Sustainable Technological Development Programme in the 1990s. His main research interests are technological innovation for sustainability, technology assessment of emerging technologies, sustainable consumption, sustainable system innovation, and small-scale experimentation and learning; with special interests in energy, housing and transportation. Together with Tellus Institute he works on the 'Great Transition Initiative' to bring about a societal transition towards sustainability. Prof. Vergragt is a co-founder and an Advisory Board member of the Greening of Industry network, and recently he co-founded SCORAI, the Sustainable Consumption Research and Action Initiative in the US and Canada. He has published more than 70 academic papers and book chapters, and co-authored two books. He obtained a PhD in chemistry from Leiden University in 1976.

Robert Wimmer has been Chairman of the independent research association GrAT (Center for appropriate Technology) since 1996. The leading idea of GrAT is to transform the ethical challenges of sustainable development into operational steps, and to initiate and realise examples of sustainable technology use. He coordinates national and international research and demonstration projects with an emphasis on: system solutions for sustainable development, sustainable building, renewable resources, renewable energy, product–service systems, cleaner production, and assessment methods. He is a guest professor at Nagoya Institute of Technology in Japan, and gives lectures at University for Applied Science in Salzburg and Wiener Neustadt, and Vienna University of Technology in Austria.

Tim Woolman is Project Coordinator at The Centre for Sustainable Design, UK, supporting international projects, small business assistance, tool development and research in eco-innovation and sustainable consumption and production. This follows research into Eco-Product Innovation and Clean Manufacturing Technologies at Warwick Manufacturing Group—developing enablers for small engineering companies to help them make innovative step changes in their environmental performance. Prior to this, Tim worked for 15 years in automotive product development with GKN and Ricardo, managing both detail design and client project delivery to cost, quality and performance targets; and recently as a Senior Design Engineer specifying components to comply with End-of-Life Vehicles legislation.

Rolf Wüstenhagen is a Director of the Institute for Economy and the Environment at the University of St Gallen (Switzerland). He has held visiting faculty positions at University of British Columbia and Copenhagen Business School. Since 2009, he has held the Good Energies Chair for Management of Renewable Energies at the University of St Gallen. His research focuses on decision-making under uncertainty by energy investors, consumers and entrepreneurs, and has been published in leading entrepreneurship, energy policy and environmental management journals. Rolf embarked on an academic career after retiring from one of the leading European energy venture capital funds.

Xiaodi Yang is a PhD Candidate at the Centre for Building Performance and Diagnostics, Carnegie Mellon University, USA.

Index

Page numbers in *italic figures* refer to illustrations